Fuels and Petrochemicals Division 2018

Core Programming Area at the 2018 AIChE Spring Meeting and 14th Global Congress on Process Safety

Orlando, Florida, USA
22 - 25 April 2018

ISBN: 978-1-5108-6447-4

Printed from e-media with permission by:

Curran Associates, Inc.
57 Morehouse Lane
Red Hook, NY 12571

Some format issues inherent in the e-media version may also appear in this print version.

Printed by Curran Associates, Inc. (2018)

For permission requests, please contact AIChE
at the address below.

AIChE
120 Wall Street, FL 23
New York, NY 10005-4020

Phone: (800) 242-4363
Fax: (203) 775-5177

www.aiche.org

Additional copies of this publication are available from:

Curran Associates, Inc.
57 Morehouse Lane
Red Hook, NY 12571 USA
Phone: 845-758-0400
Fax: 845-758-2633
Email: curran@proceedings.com
Web: www.proceedings.com

TABLE OF CONTENTS

Some abstracts in this book refer to images that were not made available by the publisher

513416 Overview of Honeywell UOP Oleflex^{TM}

Abstract:

Proven since 1990, the Honeywell UOP Oleflexâ„¢ technology uses catalytic dehydrogenation to convert propane and isobutane into valuable propylene and isobutylene, respectively. Oleflexâ„¢ has demonstrated the lowest cash cost of production and the highest return on investment compared with competing technologies. It minimizes energy consumption and air/water emissions, and uses a recyclable, platinum-alumina-based catalyst to reduce its impact on the environment. The unique process design maximizes operating flexibility and on-stream reliability. This presentation will review the Oleflex process, highlight the key design benefits, and summarize the evolution of technology enhancements since the first commercial unit nearly 30 years ago.

513835 Propane Aromatization Using Sulfated Zirconia Supported Mo Catalysts

Abstract:

Aromatization of propane into benzene and other aromatics has been a very well-studied reaction over Ga- and Zn- doped HZSM5 catalysts. In aromatization, both metal and acidic sites are required. Although Mo-doped HZSM5 is known to have both of these sites, and has been extensively studied for CH_4 aromatization, it has not been widely reported for propane aromatization. In addition, we are not aware of the use of Sulfated Zirconia (SZ) as a solid acid for propane aromatization, despite its well-known superacid properties. Like HZSM-5, SZ also possesses strong Brønsted acid sites that can catalyze olefinic intermediates into aromatic products. To our knowledge, no other reports have tested SZ supported Mo catalysts for the aromatization of propane, despite the fact that this catalyst contains both the Mo metal and acid sites required for this reaction.

In the present study, Mo was doped on SZ and these catalysts were characterized using DRIFTS, XRD. Using pyridine adsorption, DRIFTS showed small effects on acidity when Mo is added to SZ in that the peaks corresponding to Lewis (L) and Brønsted (B) acid sites are shifted to lower wavenumbers as shown in Figure 1. XRD showed no peaks for MoO_x species indicating high dispersion of Mo, and SZ crystallinity is not affected by the addition of Mo. This catalyst was then evaluated for propane aromatization around 650 ^{0}C. Primary products included ethylene, ethane, and propylene along with aromatics including benzene, toluene and xylenes. However, catalyst deactivated over time, likely due to coking on the surface of the catalysts that was confirmed by TPO. The effect of temperature and Mo content on acidity, activity, and selectivity was also investigated. Temperature plays a key role as higher temperatures although showed high activity, it also leads to faster deactivation. Whereas, lower temperatures showed lower activity but lesser rate of deactivation. Higher Mo content also increased the activity as more number of active sites are available for the intermediates to form but too high Mo can lead to non-uniform dispersion and can affect the acidity of the catalyst significantly.

Figure SEQ Figure * ARABIC 1: Comparison of acid sites for SZ and Mo doped SZ catalysts (after pyridine desorption at 100 ^{0}C, (a) Sulfated Zirconia (SZ); (b) 1%Mo doped SZ; (c) 5%Mo doped SZ; (d) 10%Mo doped SZ), L= Lewis acid site, B= Brønsted acid site.

512989 Oxy Dehydrogenation of Paraffin's to Olefins Using Co5MgAlO Catalyst

Abstract:

OXIDATIVE DEHYDROGENATION OF LIGHT PARAFFINS (C_2 â€“ C_5) INTO OLEFINS: using $Co_{(5)}$MgAlO as catalyst

Abstract

Olefins have enjoyed many uses in a wide variety of industries, from car manufacturing to energy production. Energy consuming processes of catalytic dehydrogenation of turning paraffins into olefins has been commercialised since the early 20th century, while catalytic oxydehydrogenation of paraffins to olefins is still in prototype stages. The conflict between kinetic, thermodynamic and yield constraints has delayed the commercialisation of this process. The solution to achieving the relevant process route will be exploitation of the right catalyst at moderate temperatures and pressure. Co_5MgAlO is studied under atmospheric pressure and temperatures not exceeding 650$^{â°}$C, to dehydrogenate C_2 to C_5 paraffins to olefins using oxygen as a reactant. The ratio between hydrocarbon and oxygen was kept at 1 mole below complete combustion. Mathematical models in relation to practical results where also examined.

Keywords; Paraffins, Olefins, Mathematical theoretical results, Practical experimental results, Paraffin dehydrogenation, Cobalt(5)MagnesiumAluminiumOxide catalyst.

512775 Ethanol to Infrastructure Compatible (Jet) Fuels Via Ethanol Coupling Reactions

Abstract:

In recent years, the maturation of the bioethanol industry has led to a more readily available platform for renewable fuels. However, the presence of the â €œblendwallâ€ problem, as well as the shift in focus towards infrastructure compatible fuels has resulted in a need for processes that can utilize ethanol as a building block for more valuable fuels and chemical products. Ethanol condensation is a well-known chemistry that couples alcohols through an aldol condensation process to generate higher oxygenates. The higher oxygenates can be directly used as a chemical commodity or easily converted to infrastructure compatible fuels. To demonstrate the viability of infrastructure compatible fuels production, in particular jet fuel, from the ethanol coupling pathway the higher oxygenates generated from the ethanol coupling chemistry sent through a sequence of catalytic processing steps to produce the fuel fractions. The final jet fuel fraction was then analyzed using ASTM methods to compare the fuel properties with standard jet values. For jet fuel, ASTM standard for the freeze point is < -40 °C and the ethanol coupling generated jet fraction is < -70°C, which surpasses the freeze point requirement. ASTM requirement for the density at 15°C is between 0.775 - 0.840 kg/L and the ethanol coupling generated jet fraction density value (0.789 kg/L) falls in the ASTM required range. This talk will discuss the individual catalytic steps involved in generating the infrastructure compatible fuels via ethanol coupling reactions and the final fuel properties.

512887 Purification of Tire Pyrolytic Oil (TPO) and Its Comparison with Diesel Fuel

Abstract:

Due to growing population and rapid industrialization the global demand for industrial rubber products is increasing. This increase in demand for rubber products has led to the increase in global rubber production which has devastating environmental effects. Disposal of waste rubber products is becoming an environmental challenge due to its non-bio degradability. The majority of the waste rubber product is generated from the worn out, damaged or discarded automotive tires also known as scrap tire. Approximately 1.5 billion tires are produced each year which eventually end up in the waste stream. Scrap tire consists of organic matter that is often disposed illegally in open lands, ultimately creating a big pollution hazard to environment and hygiene. Pyrolysis is considered as a useful technique for recycling of scrap tire by which oil, gas, carbon black and steel wires are obtained. Pyrolytic oil is currently receiving renewed interest as an alternative to fossil diesel fuel. In the present work, crude TPO obtained from vacuum pyrolysis method was purified via de-sulfurization followed by simple distillation. Decolorization and deodorization of the oil was also done to make it suitable for the industry. 30-35% of purified oil (product) was obtained after processing. A comparison of different properties of oil is done before and after purification. Properties of the purified TPO are also compared with diesel fuel for its suitability in diesel engines.

513260 Using Advance Control to Monitor Plant Performance and Economics

Abstract:

With thousands of applications and over a forty-year track record, multivariable predictive control has become the gold standard for advanced process control within the processing industries. Using a dynamic model of the process unit, multivariable predictive control can simultaneously optimize unit operation while maintaining strict control of key operating constraints. Yet as ubiquitous as these controllers are, they are largely under the influence of the control or advanced control engineer and provide little feedback to the process engineer or operations manager whose involvement is critical for continuous improvement of the plant.

This presentation discusses techniques for bridging this knowledge gap by providing operational insights and analytics using information obtained from the multivariable predictive controller. Included are techniques to monitor and assess lost opportunity costs, equipment constraint violations, and the plant operating envelope.

510712 Process Monitoring of Batch Bioreactor Process Using Optimized Data Based Model

Abstract:

Monitoring is an important aspect in industrial processes to ensure process safety, and maintain high product quality. Two important steps of process monitoring are fault detection and diagnosis. Data-based fault detection techniques such as multivariate statistical method have been widely discussed in the literature, and in this work, we have used partial least square (PLS), an input-output model to analyze the batch data [1]. In batch processes, the data is first arranged in a multiway fashion according to batches for easier access to train and test the PLS model. Most of industrial data have noise, correlated variables, and unknown disturbances, and thus we have introduced a filter by using multiscale based wavelet functions to de-noise, de-correlate and normalize the dataset [2-3]. Orthogonal wavelet functions decompose the original data signal into multiscale approximate, and detailed scale and additional statistical filter is applied to remove unknown disturbances, and then the filtered signals are reconstructed to form global signals.

Most of the industrial process are nonlinear in nature and application of linear PLS will lead to mismatch between model and process data, hence in this work we have used kernel extension of PLS technique as our nominal data-based model. Kernel partial least square (KPLS) is first trained with normal operating historical data to compute all the training score vectors. KPLS model is applied on the global scale to the reconstructed dataset to generate model residue or error, and fault detection decision is made by performing statistical analysis on the residue. In literature conventional statistical test, and test have been studied but they have been inadequate to diagnose the fault, and show inferior performance while detecting multiple faults compared to advanced statistical test like generalized likelihood ratio test (GLRT) [4]. GLRT solves composite hypothesis test, and provides an ability to analyze the multivariate data in time and also along individual variables, and thus improving fault detection performance for detecting multiple faults. As soon as the fault is detected, it is necessary to find the variable that presents the major deviation from its expected value; therefore, a contribution plot with GLRT is performed for diagnosis.

To further improve performance of KPLS model, selection of kernel function and its parameter is optimized, and in this work we have used multi-objective genetic algorithm to optimize selection of kernel function and its parameter value by minimizing missed detection rate and false alarm rate. KPLS being an input-output model, can also be used as nonlinear regression tool to predict the product concentration from online sensor measurements and thus providing an additional option to monitor the concentrations that are not available by sensor measurement. The developed model is applied to batch bioreactor process; experimental data of batch production of beta-carotene via fermentation is used to describe glucose consumption, metabolic product formation and depletion, and the beta-carotene production in the *Saccharomyces cerevisiae* strain mutant SM14, with 20 g/l glucose as the carbon source. Experiments are performed to validate the results obtained from the model; the process monitoring results by our developed model shows superior performance compared to conventional fault detection techniques.

Keywords: KPLS, GLRT, wavelet function, fault detection, batch processes, bioreactor.

References:

1. MacGregor, J.F., and T. Kourti. "Statistical Process Control of Multivariate Processes." Control Engineering Practice 3, no. 3 (March 1995): 403–14.
2. Bakshi, Bhavik R. "Multiscale PCA with Application to Multivariate Statistical Process Monitoring." American Institute of Chemical Engineers. AIChE Journal 44, no. 7 (July 1998): 1596.
3. Botre, Chiranjivi, Majdi Mansouri, M. Nazmul Karim, Hazem Nounou, and Mohamed Nounou. "Multiscale PLS-Based GLRT for Fault Detection of Chemical Processes." Journal of Loss Prevention in the Process Industries 46 (March 2017): 143–53
4. Mansouri, M., Nounou, M., Nounou, H., Karim, N., 2016. "Kernel pca-based glrt for nonlinear fault detection of chemical processes." Journal of Loss Prevention in the Process Industries 26 (1), 129–139.

510706 Machine Learning at Scale for the Advanced Process Control Toolkit

Abstract:

The last decade has seen an explosion of interest and development in the fields of Big Data, Big Data Analytics, and Machine Learning. Recent research has recognized that classical approaches to realizing Artificial Intelligence (AI) has its limitations, and has instead focused on learning systems based on statistical techniques. This approach has led to the rapidly expanding field of Machine Learning.

Machine Learning relies heavily on numerical optimization; this is similar to the significance of optimization algorithms in advanced process control (APC)/model predictive control (MPC) and Online Optimization. As we know, MPC relies on a Linear program (LP) and/or Quadratic program (QP) while Online Optimization typically utilizes nonlinear optimization algorithms. The fields of APC/MPC and Machine Learning have more in common than is obvious at first glance. In APC/MPC, the optimization objective is to minimize a sum of stage costs (finite or infinite); supervised machine learning, for instance, seeks to minimize a sum of stage empirical risk functions/costs.

The objective of this talk is to demonstrate how large-scale machine learning algorithms can be a useful tool in the APC practitionerâ€™s toolkit. Specific large-scale machine learning algorithms are presented with case studies. The first framework is a learning algorithm that allows easy tuning of the objective function in APC. The second framework focuses on how AI algorithms can be used to build and adapt inferentials and/or property estimation correlations. Finally, the real time asset monitoring problem is discussed. Performance guarantees of the algorithms will be briefly discussed.

DRIVING AMMONIA PLANT OPERATIONAL EXCELLENCE THROUGH PERFORMANCE MONITORING AND ANALYTICS

Satish Baliga
KBR
Houston, TX

Avinash Malhotra
KBR
Houston, TX

Abstract

As a licensor of ammonia technology, KBR has deployed a cloud-based, performance monitoring, analysis and diagnostic service at several grass-roots ammonia plants world-wide. This service utilizes operational insight for early detection of incipient problems that if ignored could lead to degradation of process and equipment performance. Smart, proactive, escalating alerts act as early warning notifications of impending abnormal or sub-optimal process deviations and the visualization platform supports quick identification and exploration of potential sources of problems.

At the heart of the service are operations experts dedicated to diagnose these problems, determine probable root cause, and recommend corrective actions that can either prevent failure or enable quick resolution of current operational problems.

This paper presents the methodology and work process in conducting in-depth analysis of process and equipment health and identifying opportunities for optimizing plant operations towards the upper limits of thermodynamic efficiency. Examples are provided based on the plant commissioning and start-up where rigorous simulation models were utilized to mimic transient behavior of the various unit operations, check out control system configuration, optimize control-loop tuning parameters and validate interlock logic related to the start-up and shutdown of the plant. These examples demonstrate the ability to remotely shadow the plant operations through a cloud-hosted monitoring and visualization platform and leverage the licensor's in-house experts to collaborate with plant personnel to reduce commissioning and start-up time and to provide long-term support for sustained targeted operations.

Introduction

Ever since the technology for the synthesis of ammonia from nitrogen and hydrogen was developed in the early 1900s, ammonia has been used as the primary intermediate for the manufacture of fertilizers. Today more than 85% of ammonia produced globally is being used to make fertilizer for food production (1). With the world population anticipated to grow by 30% in the next 30 years, the increasing demand for agricultural goods has continued to spur the construction of ammonia plants in highly populous regions of the world such as China and India as well as several developing nations in Southeast Asia. Even the U.S. where no new ammonia plants were previously built for almost twenty years has been looking to capture a lot of that demand over the past four years by leveraging the abundant supply of low-cost natural gas domestically. Today some of the older ammonia plants are being recommissioned and upgraded and several new ones have been built or are being proposed for construction (2).

An important challenge that arises in the face of this spurt in the construction of ammonia plants globally is the availability of skilled labor workforce to operate these new plants. In places like the U.S. this is further compounded by the fact that many members of the skilled labor workforce will be reaching retirement age over the coming decade. As per the ARC Advisory Group (3), one way for owner-operators to address this challenge is to engage technology licensors and third party process experts to actively monitor their plant operations as a way to provide cost-effective solutions for improving plant performance and equipment health. KBR is the world's leading supplier of ammonia process technologies, and has been involved in licensing, designing, engineering and constructing more than 230 ammonia plants accounting for approximately half the current total global ammonia production. This paper describes some common operational issues experienced during the commissioning and startup of an ammonia plant and illustrates how a technology licensor like KBR can leverage in-house expertise to assume the role of a trusted advisor and provide timely guidance to plant personnel over different phases of plant operations.

Process overview

Ammonia production is an energy intensive process consisting primarily of steam reforming, shift reactions, methanation, CO_2 absorption, and ammonia synthesis. The type of feedstock has a fairly significant impact on the amount of energy that is consumed and carbon dioxide (CO_2) produced. About 70% of global ammonia production is based on steam reforming of natural gas which is considered the most efficient technology with respect to energy use and CO_2 emissions. Relatively speaking, ammonia production is fairly simple. The inputs are nitrogen, hydrogen and heat. The nitrogen used in the process is taken from the air and hydrogen is obtained through steam reforming of natural gas. The main reactions involved in the production of ammonia are as follows:

Primary Reforming

$CH_4 + H_2O \leftrightarrow 3H_2 + CO$ (endothermic reforming)

$CO + H_2O \leftrightarrow CO_2 + H_2$ (exothermic shift)

The overall reforming reaction is highly endothermic and requires heat from natural gas fuel.

Secondary Reforming

$CH_4 + H_2O \leftrightarrow 3H_2 + CO$ (endothermic reforming)

$2H_2 + O_2 \rightarrow 2H_2O$ (exothermic combustion)

$2CH_4 + 3O_2 \rightarrow 4H_2O + 2CO$ (exothermic combustion)

$CO + H_2O \leftrightarrow CO_2 + 3H_2$ (exothermic shift)

The heat required for the reforming reaction in the secondary reformer is supplied by the heat of combustion of hydrogen and methane with oxygen.

Ammonia Synthesis

$N2 + 3H_2 \leftrightarrow 2NH_3$ (exothermic)

KBR Purifier™ Ammonia Process

A block flow diagram of KBR's Purifier process for ammonia production is shown in **Figure 1**.

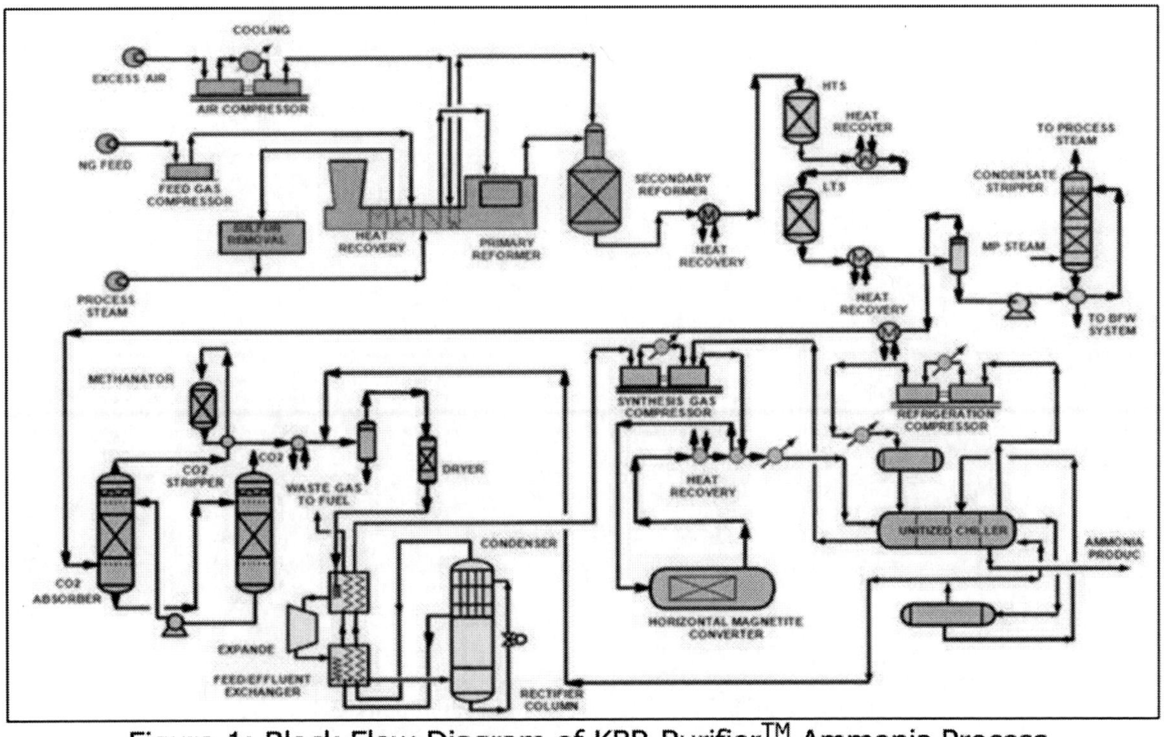

Figure 1: Block Flow Diagram of KBR Purifier™ Ammonia Process

In KBR's Purifier™ ammonia process, air flow to the secondary reformer is set to achieve a hydrogen-to-nitrogen ratio (H/N) of 2:1 in the feed gas to the Purifier. About 30% of the reforming is shifted from the primary to the secondary reformer. This results in a much lower temperature (690-715 °C) of the process gas exiting the primary reformer tubes as well as a lower temperature (880-900 °C) at the exit of the secondary reformer. The Purifier provides the makeup gas for the synthesis section with a hydrogen-to-nitrogen ratio of 3:1 and around 0.25-0.30% of inert gases including argon. Most of the unreacted methane, excess nitrogen and some argon is removed in the waste gas from the Purifier and used as secondary fuel in the primary reformer.

The energy efficiency (4) of the steam reforming process is about 28–35 GJ per tonne of ammonia. Most plants operate above the ideal, thermodynamic minimum energy consumption with top quartile performers ranging between 28 and 33 GJ per tonne. In KBR's Purifier™ process, the low temperature of the primary reformer results in significant energy savings and makes the Purifier process one of the most energy efficient technologies for producing ammonia today.

Performance Monitoring and Analytics

KBR has deployed Ammonia InSite™ a cloud-based performance analysis and diagnostic service at several grass-roots ammonia plants world-wide. Ammonia InSite™ uses a combination of state-of-the-art web-enabled visualization dashboards and advanced mathematical algorithms to leverage and unlock the value of near real-time plant data in order to create a visual "digital fingerprint" of the plant. An example of dashboard that provides a quick overview of the current status of an ammonia plant is shown in **Figure 2**.

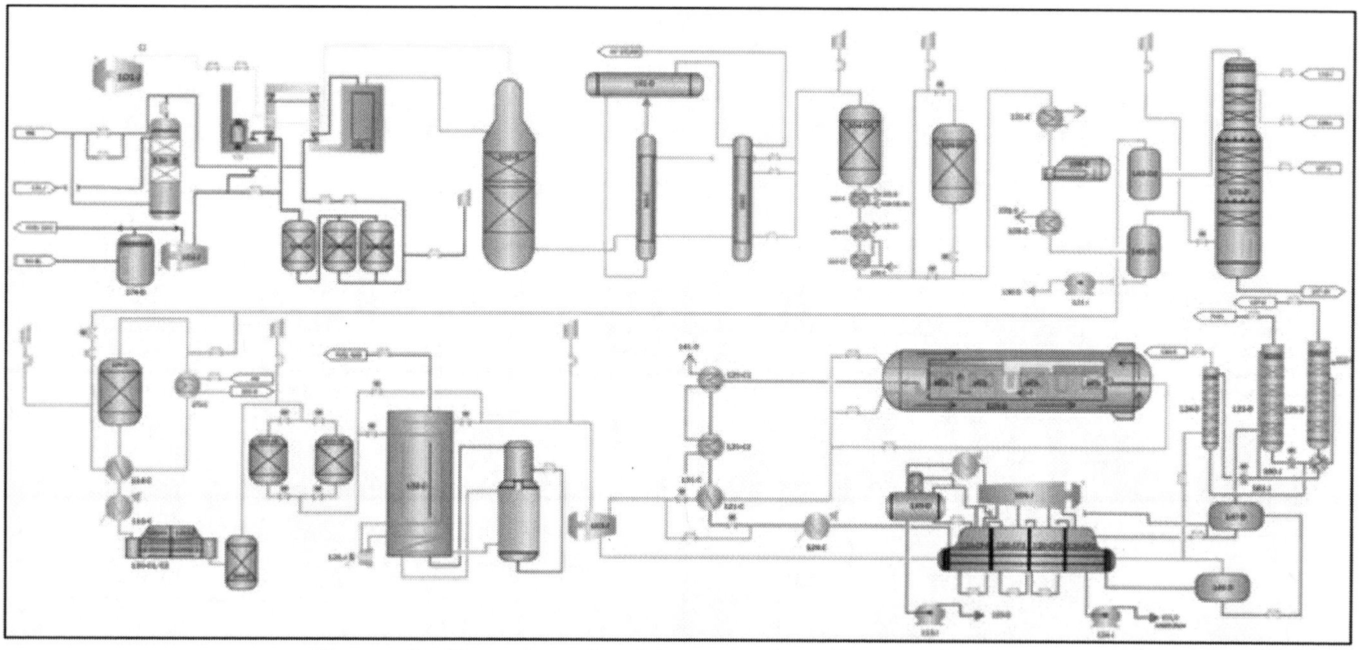

Figure 2: Overview Dashboard of an Ammonia Plant

Since these dashboards are cloud based and web-enabled, they are available 24x7 to KBR experts as well as to authorized plant personnel so that both the service provider and the customer have the same identical view of the plant performance.

At the core of this service is the availability of a pool of KBR process and equipment experts who leverage their in-depth technical know-how of the ammonia process to provide timely guidance to plant personnel to support improvements in plant operations and equipment reliability. The process experts use the dashboards as an enabling platform to visualize and diagnose operational problems, determine probable root cause, and recommend corrective actions that can either prevent abnormal operations or enable quick resolution of current problems. Smart, proactive, escalating alerts act as early warning notifications of impending abnormal or sub-optimal process deviations. Drill-down features allow the rapid visual identification and exploration of potential problem sources.

Proprietary calculation methods and first-principle-based simulation models that utilize real-time data from the plant provide the necessary framework to conduct in-depth analysis and predict the current process performance and the health of associated equipment. Inherent or derived parameters such as exchanger heat transfer coefficients, compressor efficiencies, specific-energy consumption, approach to equilibrium, etc. are examples of parameters calculated periodically by these first principle models.

In order to maintain the integrity and validity of these calculated parameters it is important to note that the plant measurements being monitored are reconciled for heat and material balance and any measurements in gross error are flagged and brought to the attention of the plant personnel. Offline, what-if scenario studies are also used by KBR's process experts to identify opportunities for production increase or energy reduction and to drive plant operations towards the upper limits of thermodynamic efficiency.

Technology Licensor as a Trusted Advisor

From KBR's experience, the opportunity to provide "Trusted Advisor" mentorship to ammonia plant personnel has been possible in the following three areas:

- Commissioning and Startup - support of grassroots ammonia plant personnel during commissioning and startup activities
- Normal steady state plant operations – using proprietary mathematical tools and subject matter expertise to transition the ammonia plant from a current steady state operation to a more optimal one
- Unscheduled plant shutdowns – conducting root cause analysis to determine reasons for plant trips and, if appropriate, redesigning interlock logic to mitigate future occurrences

Each of these areas is discussed below in detail along with relevant use cases.

Commissioning and Startup

KBR has commissioned nine Purifier™ based ammonia plants (5) over the past three years and several more are scheduled for start-up early this year. While the sizes of these plants vary from 1000 to 2900 metric tons per day, they share many common features with each plant design accommodating specific site requirements of the individual customers. Successful commissioning of these plants requires careful planning and utilization of lessons learned from many successful start-ups.

As is typical with turn-key projects, KBR as the detailed engineering contractor has provided commissioning and start-up teams for onsite service while being supported by technology advisors from the home office. To a large extent, the deployment of the Ammonia InSite™ service on these projects has increasingly contributed to the safe and successful start-up of these plants. Specifically, the ability to remotely monitor plant operations in near real-time through the visualization dashboards has allowed KBR experts located at different engineering offices to collaboratively shadow commissioning and startup activities and alert the onsite team of any deviations from best practice guidelines.

A few examples of use cases for Ammonia InSite™ service during the commissioning and startup phase are listed below.

Rate-of-change (ROC) alerts: ROC alerts are typically configured on the Ammonia InSite™ dashboards for key variables being monitored during startup or shutdown of major process units. At any time, if the ROC exceeds limits specified in the standard commissioning guidelines, alerts and e-mail notifications are automatically dispatched to process engineers in the home office who are assigned to remotely support onsite activities. The process engineers then engage with onsite personnel to discuss remedial action if required. Examples of ROC alerts setup during the commissioning and startup phase include the following:

a) rate of change of reformer outlet temperature (high ROC can induce thermal stress in the catalyst tubes)
b) rate of change of pressure at the inlet of the ammonia reactor (high ROC can disturb the catalyst beds)
c) rate of change in inventory levels in columns (high ROC can dislodge column internals or lead to loss of liquid seal)

Instrument Calibration: With remote access to real-time data, process engineers in the home office are able to collaborate with control system engineers at the site to validate instrumentation readings and identify sensors that have to be repaired or require recalibration. Much of this validation is done in the home office through the use of dynamic simulation programs which are developed to closely mimic transient conditions normally observed during startup of the various process units. In some cases, such validation has identified improper pressure-temperature and/or molecular weight compensation of flow measurements. Similarly, unusual compositions reported by online process gas analyzers due to improper calibration are also detected and rectified.

A calculation error that has sometimes been observed is related to the reformer steam-to-carbon ratio which is a very critical operating parameter in an ammonia plant. Errors in the measurement of this variable have a strong impact on energy consumption as well as on equipment reliability. In one instance, poor calibration in the steam flow measurement resulted in a much lower actual steam-to-carbon ratio than indicated on the operators' control system panel. The problem was identified by the process expert remotely monitoring the plant who recognized an inconsistency between the temperature rise across the HTS converter and the inlet and outlet CO measurements. Fortunately, since steam-to-carbon ratios are normally kept relatively high during the early phase of start-up, the issue was addressed in time and no damage was done to the reformer catalyst or tubes.

<u>Control Loop Tuning</u>: One aspect of commissioning that is quite often understated is the tuning of the base control loops in an ammonia plant. Experience has shown that neglecting to invest sufficient time and effort in the tuning of loops has a long term impact on the efficient operation of a plant. Quite often, onsite personnel with insufficient experience or training with control loop tuning adopt a trail-and-error approach which is beset with risks. To mitigate these risks and prevent unstable behavior of key process variables, the approach recommended and practiced by KBR is to use a training simulator or a dynamic simulator to generate and test out control loop tuning constants before they are deployed on the real process.

For key loops the set points, process variables and controller outputs can be included in the scope of monitoring so that the impact of poorly tuned loops on process variability and stability can be easily studied. For instance, the ability to safely push a plant towards its constraints is limited by the stability of the process. When it comes to pushing the plant towards the upper limits of production, operators tend to be more conservative when key process variables are susceptible to disturbances due to poorly tuned control loops.

Optimization of steady state plant operations

Continuously improving normal steady state operations of a plant through timely expert advice based on proactive monitoring and data analysis is a primary benefit derived from the Ammonia InSite^{TM} service. Examples of use cases across several ammonia plants that KBR has commissioned and started up during the past couple of years are highlighted below:

<u>Vent Loss Minimization:</u> Producing more ammonia for a given amount of natural gas is easily the most cost effective way to improve the plant's operating profits. To achieve this objective, all process gas vents across the ammonia plant are closely monitored though the Ammonia InSite^{TM} dashboards. Using individual valve characteristic equations, the flow through these valves (when open) is estimated and then aggregated over the month. **Figure 3** shows the totalized flows through these vents from a representative plant. These vent losses either cause direct loss of ammonia production or in the case of waste gas vents result in additional import of fresh natural gas fuel.

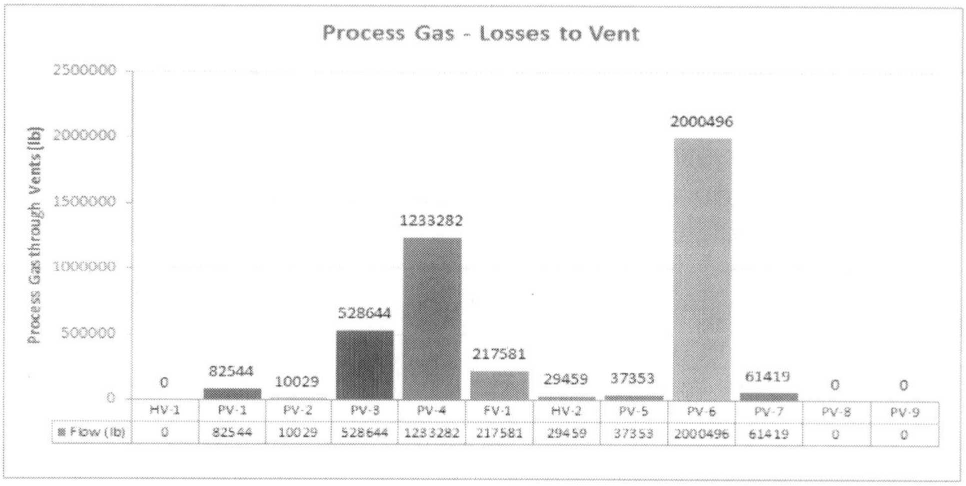

Figure 3: Loss from Process Gas Vents

In the specific case shown in the figure, a total of 1875 metric tons of process and waste gas was vented and contributed to a 1% potential loss in ammonia production for the month. Under normal non-flaring situations, these vents are opened by plant personnel only to counteract some operational issues. The role of the trusted advisor would then be to identify the root cause of these issues, find alternate ways to address them, and to the extent possible close the vents to minimize production loss.

Ammonia Reactor Optimization: A truly effective advisory service addresses both real-time operational issues (as in the case of open process vents described above) as well as longer term improvements achieved through on-demand process optimization. A great example of this described below is the inlet temperature optimization in a multi-bed ammonia reactor that would help sustain conversion over the long term even as the catalyst activity slowly degrades with time.

The reaction between hydrogen and nitrogen to form ammonia is a reversible exothermic reaction. Hence an increase in temperature results in a decrease in the equilibrium constant and a subsequent decrease in conversion. Also the tremendous amount of heat evolved during the reaction increases the temperature of the catalyst bed and if too high it can permanently damage the catalysts. In order to avoid this situation, KBR's ammonia converters utilize inter-bed cooling systems, which remove the heat of reaction. Lower inlet temperatures however result in lower reaction rates. The challenge therefore is to maintain the inlet temperature at levels that result in high conversions at acceptable reaction rates. These inlet temperatures can be adjusted by manually manipulating hand-valves on the intercooler inlet streams and the intercooler bypass to maintain the bed temperature as close to the optimum target as possible.

In the specific example shown in **Figure 4**, by reducing the inlet temperature of the second and third bed of the reactor the ammonia exit concentration was increased by nearly one mole%.

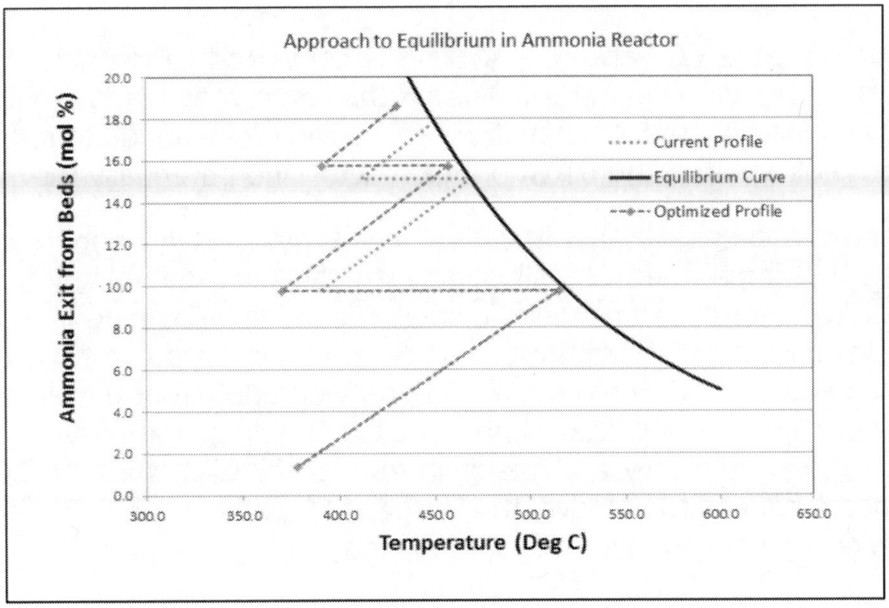

Figure 4: Optimizing Ammonia Reactor Equilibrium Operations

Even though the decision to lower the bed inlet temperatures to improve conversion was reached after several weeks of tracking the reactor performance, it can still be categorized as a proactive outcome of an advisory service that relies on continuous monitoring and analysis as the prerequisite for plant performance improvement.

CO_2 Removal System foaming: The function of the CO_2 removal system in an ammonia plant is to remove CO_2 produced by the shift reactions occurring in the frontend of the plant. KBR uses the BASF Two-Stage OASE System shown in **Figure 5** to absorb CO_2 from the process stream and limit its content to 500 ppm or less.

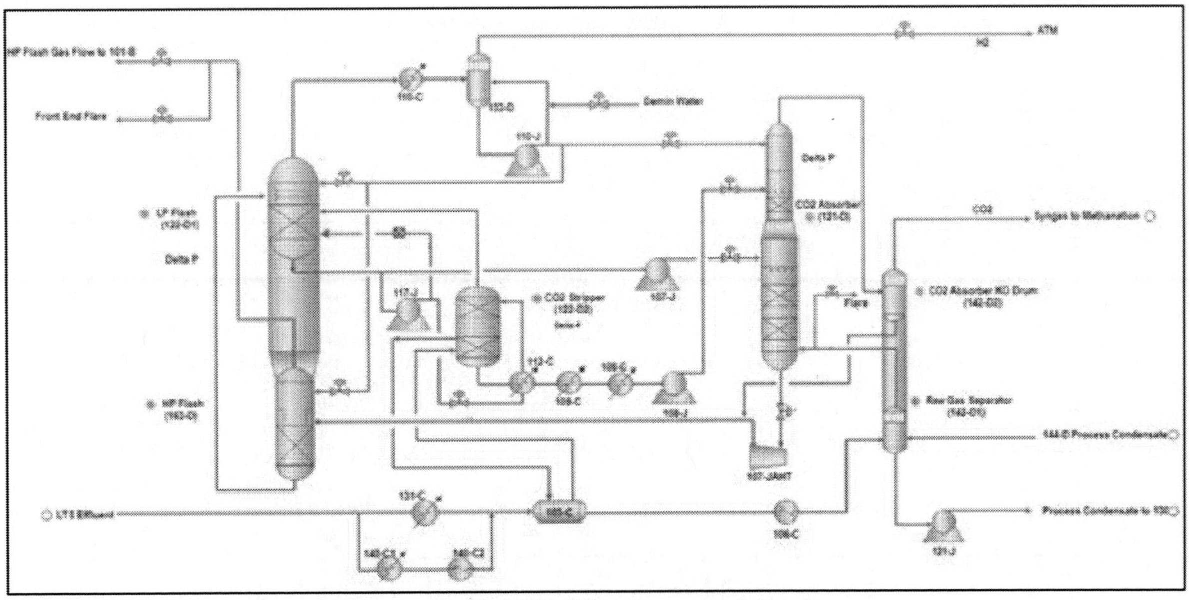

Figure 5: OASE CO_2 Removal System

Any CO_2 leaving the CO_2 absorber system is converted to methane in the Methanator downstream of the CO_2 removal system. The methanation reaction is highly exothermic and high levels of CO_2 entering the Methanator can cause rapid outlet temperature excursions which in many cases leads to a shutdown of this reactor and potentially the whole plant.

It is imperative therefore that the CO_2 levels are closely monitored along with the process variables that impact the CO_2 slippage, namely, the OASE circulation rates, absorber temperatures, as well as a phenomenon known as "Foaming". Foaming is typically caused due to the presence of fine particles (catalyst dust or corrosion products from the piping) in the circulating solution used for CO_2 absorption. Gradual degradation of this amine solution due to continuous operation of the unit also contributes to foaming. Excessive foaming can cause reduction in absorption efficiency and result in sudden breakthrough of CO_2. Maintaining a continuous filtration system to remove the suspended particles and/or periodic addition of anti-foaming dosing agents can help mitigate foaming.

Foaming is not a directly measured phenomenon. However, increase in pressure drops in the flash columns and/or the absorber columns can typically be used as early indicators of excessive levels of foaming. The trick however is to continuously monitor these pressure drops and be in a position to predict into the future whether foaming could become an issue and potentially cause CO_2 breakthrough into the Methantor. **Figure 6** shows a data-driven model that utilizes the flash column pressure drop as a leading indicator to predict with almost 95% certainty when foaming could become a significant issue. The results though preliminary are promising and validate the applicability of big data methods to complement the well-established first-principle modeling approach to data analysis.

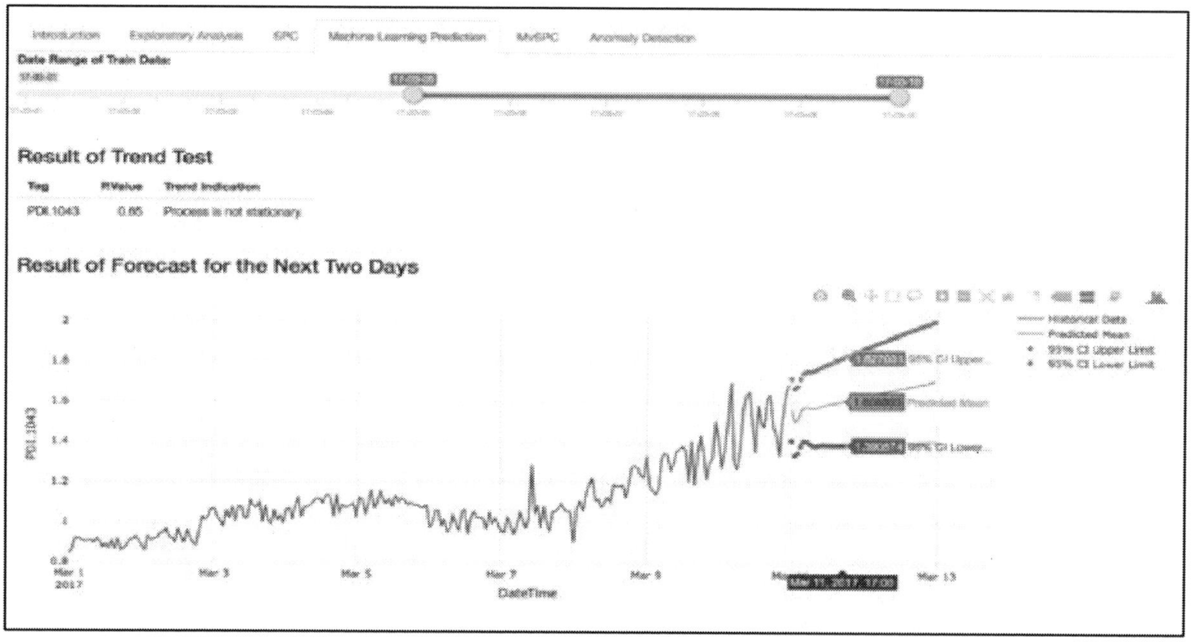

Figure 6: Big Data Predictive Models to predict foaming

Overall Heat and material balance based Reconciliation: As part of the Ammonia InSite™ service, KBR uses high-fidelity simulation models of the ammonia plant to conduct an overall material and energy balances based on thermodynamics, fluid mechanics and material transfer equations. The benefit of this approach is that while closing the heat and material balance, the simulation models update intrinsic parameters, such as efficiency, fouling, catalyst activity and such while simultaneously reconciling plant measurements in the least square sense, taking into account the inherent accuracy of sensors. Two uses cases described below justify the effort required to build these models and to include it in the scope of a technical advisory service such as Ammonia InSite™.

As indicated earlier, the steam-to-carbon ratio impacts production, energy usage as well as the reliability of the Primary Reformer. Too low a ratio can result in carbon deposits on the catalyst which increases the resistance to gas flow in the primary reformer tubes and may impair catalyst activity. A high ratio in excess of what is recommended by KBR's process experts results in a high overall energy usage. It is therefore important to closely monitor the steam-to-carbon ratio, reconcile the relevant measurements, and make the adjustments necessary to maintain a good balance between production and reliability. A specific example from a customer site is the reconciliation of the steam-to-carbon ratio from the measured value of "X" to a reconciled value of "X + 0.3". This implies that the plant was actually operating with a significantly higher Steam-to-Carbon ratio than what was measured. Gradually lowering the plant measured value by 0.3 would increase the steam available for export by at least 10 tons/hour resulting in improved profitability.

Another special case of a heat and material balance based reconciliation conducted on an ammonia plant involves the validation of an OSBL natural gas custody flow meter that did not match very well with the flow measurement located within the ISBL. This had direct financial implications because the plant operator was being invoiced for the natural gas feed imported into the plant by the gas supplier. The results of the data reconciliation indicated that the OSBL flow measurement for natural gas import was actually more accurate than the ISBL flow measurement and the customer wad advised to recalibrate the ISBL measurement appropriately.

Unscheduled plant shutdown

In brownfield plants most unplanned shutdown are directly caused by equipment failure (6) usually as a consequence of regular wear and tear associated with the operation of such equipment. On the other hand, at grass root plants many of the unplanned shutdowns are attributed to process upsets, failed sensors, abnormal weather conditions, errors in trip-logic, and in some situations due to operational oversight of inexperienced operators.

In all cases, the design intent is to isolate (by flaring through vent valves) the shutdown to just those sub-areas of the plant that are directly impacted and to maintain the rest of the ammonia plant in a stable condition in order to facilitate a quick restart of the plant if possible.

However, this is easier said than done given that the frontend and backend of the plant are tightly heat integrated through the steam system. Specifically, the main compression systems in the backend (synthesis gas and refrigeration) are typically driven by turbines that utilize steam generated through the recovery of process heat by waste-heat boilers in the frontend. So more often than not with the exception of plants run by experienced operators, the effect of a steam turbine trip in the backend cascades very quickly to a frontend trip.

Playing the part of a trusted advisor, KBR has resolved the cascading trip scenarios at several plants using a rigorous dynamic model of the plant. The interlock logic that gets triggered during such trip events has to take into consideration dynamic aspects of back pressure, speed of response of letdown valves between the high and medium pressure steam headers, tuning of steam header pressure controllers, rate of change of plant load at the onset of the trip event and other such factors. The use of a well-tuned dynamic simulator to mimic the sequence of events that precede the trip, allows the root cause of the trip to be analyzed, controllers to be retuned, and the interlock logic to be modified and thoroughly tested to account for relevant process dynamics as well as operator action. Without the Ammonia InSite™ service, plant operator would have had to learn from each event after the fact and plan proper anticipatory response to address similar events in the future. This "learn from experience" approach is not ideal, could be expensive depending on how effectively it improves plant uptime, and could be counter-productive if the root cause is not fully diagnosed. On the other hand, KBR's experience (7) indicates that a dynamic simulation based trip analysis can reduce the number of unscheduled cascade trips and increase the plant uptime by 3%.

Summary

Operating plants are often too resource constrained to manually inspect and interpret process and equipment data at a frequency required to meet reliability goals. A three-pronged approach for achieving operational excellence involves simultaneously driving plant operations towards production maximization, energy minimization and extended equipment uptime. Continuous visibility into the plant's current as well as historical operating profile provides a look-back as well as a predictive assessment of asset performance and uncovers opportunities for sustained efficiency gains. Close collaboration with technology licensors enables plant operators to leverage the domain knowledge of their process technology experts and helps them make informed, long-term operation and business decisions.

References

1. P. Heffer and M. Prud'homme, "Fertilizer Outlook 2013–2017", 81st IFA Annual Conference, Chicago, USA, May 20–22, 2013
2. Celeste LeCompte, "Fertilizer Plants Spring Up to Take Advantage of U.S.'s Cheap Natural Gas", SCIENTIFIC AMERICAN, April 25, 2013
3. Larry O'Brien , "Optimizing Plant Performance through Process Knowledge", ARC View, by on April 24, 2014

4. "Ammonia and the Fertilizer Industry: The Development of Ammonia at Billingham Johnson Matthey Technol. Rev.", 2018, 62, (1), 32–47
5. George Colman, Robert Collins, James Price, "Commissioning of world scale KBR Purifier™ ammonia plants", Nitrogen + Syngas 2017 International Conference & Exhibition (London 27 February-2 March 2017), pp 113-121
6. Manish Goswami, Ankita Pandey S. Nand,"Causes for Loss Time in Ammonia-Urea Plants", Indian J. Fert., Vol. 11 (2), pp.16-22 (7 pages)
7. Neda Felorzabihi and Jeffrey Feng, "Is Steam System Instability Affecting Plant Performance", American Fuel & Petrochemical Manufacturers 116th Annual Meeting, March 13, 2018

513241 A Review of Generalized Likelihood Ratio Based Monitoring Techniques

Abstract:

Fault detection is an integral part of process monitoring, which is essential for the safe and economical operation of many processes. Statistical hypothesis testing methods, such as the Generalized Likelihood Ratio chart have been utilized in order to provide improved detection results, by using the concepts of likelihood ratio and maximum estimation of their hyper-parameters in order to accurately distinguish between normal and faulty operating conditions [1,2].

Unfortunately, most of the currently available fault detection methods in literature utilize a GLR chart designed specifically to detect only shifts in the mean. However, in practice an industrial process can experience a wide variety of fault types, i.e., shifts in the variance, and drifts. The GLR charts designed specifically for different scenarios, i.e., to detect shifts in the mean, and to detect shifts in the variance, have shown that they are capable of efficiently detecting the different classes of faults, and that they provide extremely low missed detection rates, and out-of-control average run length (ARL_1) values even for small fault magnitudes [3,4].

Exponential Likelihood Ratio (ELR) charts were developed in order to enhance the performance of the GLR chart by incorporating exponential weights in order to obtain a more accurate estimate of the current state of the process [5]. Like the GLR charts, the ELR charts can also be designed to detect different classes of faults. However, the performance of the ELR charts are tied to the choice of its smoothing parameter, and therefore one objective of this work is to provide an algorithm to select the appropriate value of the smoothing parameter for monitoring purposes.

Other exponentially weighted GLR techniques exist [6]. Unfortunately a comparison and review of the performance of all likelihood-based monitoring techniques is unavailable in literature. Therefore, another objective of this work is to compare the performance of all likelihood-based monitoring techniques, in order to conclude which chart provides the best monitoring performance. Illustrative examples, such as the Tennessee Eastman Process will be utilized in order to evaluate the performances of the different techniques.

References

[1] M.R. Reynolds, J.Y. Lou, An Evaluation of a GLR Control Chart for Monitoring the Process Mean, J. Qual. Technol. 42 (2010) 287–310.

[2] D.C. Montgomery, G.C. Runger, Applied Statistics and Probability for Engineers, 5th ed., John Wiley & Sons, Inc., Hoboken, NJ, 2011.

[3] M.R. Reynolds Jr., J. Lou, A GLR control chart for monitoring the process variance, in: W.P. Lenz HJ., Schmid W. (Ed.), Front. Stat. Qual. Control 10, 2012: pp. 3–17. doi:https://doi.org/10.1007/978-3-7908-2846-7.

[4] M.R. Reynolds, J. Lou, J. Lee, S.A.I. Wang, The Design of GLR Control Charts for Monitoring the Process Mean and Variance, 45 (2013) 34–60.

[5] J. Zhang, C. Zou, Z. Wang, A control chart based on likelihood ratio test for monitoring process mean and variability, Qual. Reliab. Eng. Int. 26 (2010) 63–73. doi:10.1002/qre.1036.

[6] M.-F. Harkat, M. Mansouri, M. Nounou, H. Nounou, Enhanced data validation strategy of air quality monitoring network, Environ. Res. 160 (2018). doi:https://doi.org/10.1016/j.envres.2017.09.023.

513813 The Journey Towards 2030, Was It to Ambitious? Update on the Impact of 2030 Vision and 2020 Transformation Program on Energy, Infrastructure and Other Projects in the Region

Abstract:

Saudi Arabia and other Gulf countries announced their vision for 2030 trying to have better future for next generations and prepare for the after oil ERA.

However, we have seen some countries as well as the World Bank calling for easing the transformation to avoid falling in a recession. In this paper we will see the effect of this vision and transformation program and other programs on the projects in the region during the past period and see how this effect the oil would and gas business and projects in the region? Will this mean more projects for the energy sector and petrochemicals or are countries moving away from there? Will low oil prices and budget deficits postpone or otherwise accelerate this transformation? If yes, then for which projects? Will we see more fuel expansion projects? Renewables? Solar? Will entertainment, infrastructure and housing projects have the highest priority?

512406 Refinery-Petrochemical Integration: Organization and Culture

Abstract:

Petrochemical plants and refineries both process hydrocarbon derived materials. Refineries focus on fuels production. Refinery products include LPG (liquefied petroleum gas), gasoline (petrol), kerosene and jet fuel, diesel, fuel oils, and coke. Petrochemicals focus on creating intermediates used to produce industrial and consumer products. Major petrochemical processes include steam cracking to produce ethylene (ethane), propylene (propene), and butadiene and aromatics plants producing benzene, toluene, xylene and other aromatics.

At intermediate points in their processes, refineries produce molecules of interest to petrochemical operations, and petrochemical plants produce molecules important to refineries. The easiest molecule to make is the one you already have. Refinery-petrochemical integration works on maximizing the return on the entire hydrocarbon value chain.

In refinery-petrochemical integration three different organizations are often present;

- refinery
- petrochemical plant
- transportation link or market maker between them

This presentation focuses on cultural and organizational differences between different entities and how that can make or break projects.

513278 One Stone Two Birds: CO_2 to Methanol and Petrochemicals

Abstract:

One stone two birds: CO_2 to Methanol and Petrochemicals.

Oil prices are making gasoline and naphtha production less attractive, and it is expected a net reduction in the long-term demand since new environmental friendly energy sources are becoming available and seem to be most attractive for the future generations. On another hand, the boom of Shale Gas set the path for the construction of new ethane crackers or the switch of flexible crackers to ethane, hence the available capacity of two main sub products from naphtha crackers and refineries will be substantially reduced, propylene and aromatics.

Technologies are available or evolving to fill this gap (i.e. Propane Dehydrogenation, Olefins Metathesis, Acetic Acid Hydrogenation) however these are not enough attractive by itself, an additional driver is needed to make new technology profitable and interesting; such is the case of Methanol. In recent years this product has been in the spotlight as a viable option to produce olefins and aromatics, i.e. in China where fuel resources come mainly from coal instead of oil and gas.

Improvement of catalysis to increase the selectivity to olefins or aromatics is making a great progress, but these processes are very demanding in terms of energy (i.e. better selectivity at high temperature or energy required for further separation.), this energy demand can be indirectly translated in a higher amount of CO_2 emissions which may become an obstacle for a further development of this alternative.

However, the chemical simplicity of Methanol makes it a flexible component, which can be produced from either fossil fuels or synthetic sources such as CO_2. Catalytic conversion of CO2 into Methanol using hydrogen is getting more attention and if this is feasible, the resulting synergy could be a simple solution for two different problems, the contribution for a sustained supply of aromatics and olefins with a positive environmental impact.

This paper presents pros and cons of this integration in terms of economic feasibility, environmental benefit.

513873 Development of a Two-Stage Thermal Deoxygenation Process for Fuels and Chemicals from Biomass

Abstract:

Thermal DeOxygenation (TDO) is the thermal decomposition of mixed salts of levulinic and formic acids that can produce hydrocarbon oils at yields approaching 80% of theoretical. These acids can be sustainably produced in stoichiometric quantities by acid hydrolysis and dehydration (AHDH) of cellulose. The TDO reaction was scaled up to a 50 L semi-batch reactor to produce more than 20 L of oil for fractionation, blending and hydrotreating. The products were thus demonstrated as suitable blendstocks for gasoline, diesel, jet fuel, and fuel oils. While scale-up facilitated kilogram quantities of products for testing, the carbon loss to char increased from 8.5% at the bench scale to 48% at 50 L floor scale. Isothermal reaction experiments have identified two reaction regimes: 1) a low temperature addition-condensation regime that is diffusion-controlled and 2) a high temperature ketonic decarboxylation regime that is reaction-controlled. Through this understanding, a two stage reaction process has achieved yields in the floor scale reactor similar to early bench scale results. Current process development efforts involve testing equipment to move from semi-batch to continuous operations and scaling up the TDO process to match the output of UMaineâ€™s one-ton-per-day AHDH pilot plant. The combined AHDH-TDO process has the potential to produce a suite of bio-based chemicals in addition to fuels, so additional results include techno-economic analyses which consider the potential for co-product revenue enhancement of renewable fuels.

513289 Development of Green Pathways for the Extraction of Medically Active Components Form Grape Waste

Abstract:

Medical sector growth continues to advance thanks in part to a boom in the biomedical industry. Increased technological advancement of this level has afforded people health and wellness not seen in previous generations leading to longer more productive lives. Growth in the Biomedical can be amenably paired with green technologies to further the field in an environmentally friendly way. Innovation in the use of natural components in health fields as a way of improving health is a most promising direction but is one existing with many bottlenecks. Despite the challenges, these technology shifts are necessary to stay in line with environmental regulation, FDA regulation, and public opinion.

The use of Supercritical CO_2 is growing more and more common in many fields. It can be used singularly, with co-solvent, or as part of more complex systems to process Biomass in a more environmentally friendly way. Waste biomass can be especially useful here. What was once not worth processing and was relegated trash or burning can be processed to obtain high valuable compounds. Grape waste from wineries and juicing facilities consists of seeds, stems, rings, meat, and leaves. Most of this waste has little to no value and is relegated to trash or animal feed. With some creating processing, this waste can be a source of valuable oil as well as high-value antioxidants, polyphenols, and flavanoids. Scale is important in these operations and thus a pilot study is being conducted on this process to find the range of capacities that are profitable for this endeavor.

TUNING PID CONTROLLERS FOR COMPLEX INTEGRATING PROCESS DYNAMICS

James Beall

Emerson Automation Solutions, Austin, TX

Biography

James Beall is a Principal Process Control Consultant at Emerson Automation Solutions with over 36 years of experience in process control. He graduated from Texas A&M University with a Bachelor of Science in Electrical Engineering in 1981 and worked for Eastman Chemical Company from 1981 until 2001 at which time he joined Emerson. His areas of expertise include process instrumentation, control valve performance, control strategy analysis and design, advanced regulatory control and multivariable, model predictive control. He has designed and implemented process control improvement projects in the chemical, refinery, pulp and paper, power, pipeline, gas and oil, and pharmaceutical industries.

Paper ID 513646

Prepared for Presentation at American Institute of

Chemical Engineers 2018 Spring Meeting

14th Global Congress on Process Safety

Orlando, FL April 22-26, 2018

Keywords: Control, PID, tuning, integrating, integrator, lag, lambda.

Abstract

Tuning PID controllers for integrating processes, such as levels, can be frustrating and counterintuitive. For example, reducing the controller proportional gain on an oscillating loop can increase the oscillation amplitude! In addition to the basic integrating response, two other more complex integrating type responses are often encountered in many manufacturing processes. These are "integrator plus dead time" and "integrator plus dead time and lag". Examples of a "dead time plus integrator plus lag" response include boiler drum level, de-aerator level, large volume reactor temperature, and large volume gas pressure. This paper will show how to recognize these types of integrating responses, measure the process dynamics and provide a straightforward method to calculate PID tuning parameters. Most importantly, this method can be used to calculate controller tuning to achieve the control response that optimizes the process performance. For example, in some cases the best process performance requires aggressive tuning with good load regulation capability. In another case it might require the "least aggressive" tuning, that maintains a process variable (e.g. level) within the maximum allowable deviation from set point, yet provides maximum attenuation of process variability. This tuning method can also be applied to lag dominant or "near integrating" processes which greatly reduces the time to measure process dynamics and can provide superior load regulation. Plant examples will be used to illustrate the concepts and results.

Introduction

The two most common categories of process responses in industrial manufacturing processes are self-regulating and integrating. A self-regulating response is characterized by a response of the process variable that stabilizes, or "self-regulates", to a new value sometime after the input to the process is moved to a new value. From the standpoint of a "proportional, integral, derivative" (PID) process controller, the output of the PID controller is the input stimulus to the process and the process variable is the input to the PID controller, also called the process variable. Figure 1 compares the response of the process variable to a step change of the PID controller output for a self-regulating process and for an integrating response. Both responses include dead time. Note that the self-regulating process PV stabilizes or "self-regulates" to a new steady state value while the PV of the integrating process continues to change. The response of the latter is an "integration" of the difference between the process inputs and its outputs, hence the term "integrating process".

Level processes typically have an integrating response, the likely exception being when the outflow of the vessel is gravity driven. Other processes can have an integrating response. A "low pressure, large volume" gas pressure control application typically has an integrating process. Another example of an integrating process is a reactor temperature controller that cascades to a "jacket temperature difference" controller. The "jacket water inlet temperature difference" controller controls the difference between the reactor contents temperature and the jacket water inlet temperature based on the set point specified by the output of the reactor contents temperature controller. Figure 2 shows these examples of integrating processes.

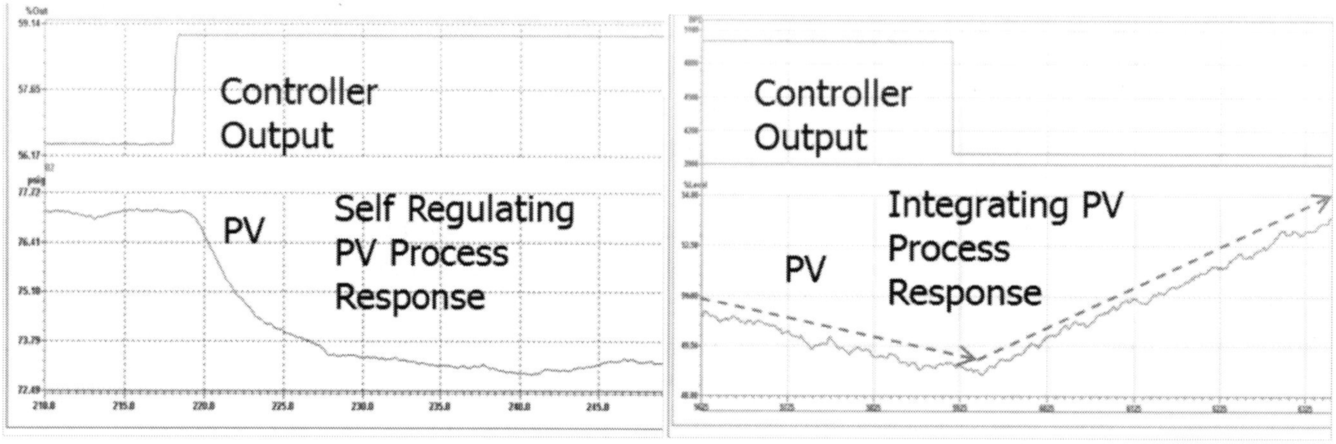

Figure 1 Response of the PV to a Step Change of the Controller Output for a Self-Regulating and an Integrating Process

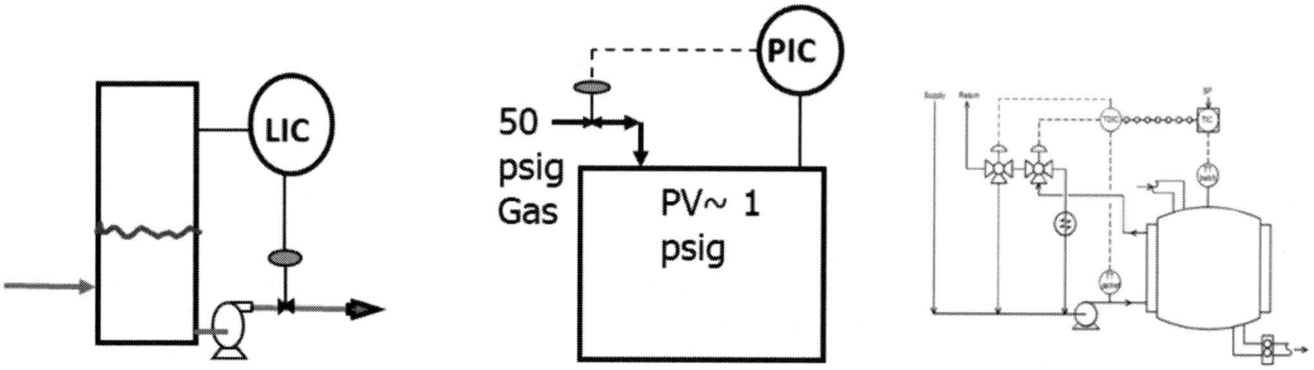

Figure 2. Three examples of integrating processes, Level, Gas Pressure and Temperature

Challenges

One of the challenges of tuning a PID controller for an integrating process is that when the integral action of the controller is combined with the integrator function of the process, the control loop will oscillate if the integral action of the controller is "too fast" (i.e. the integral time is too short). It is not intuitive to know when the integral time is too short. Another challenge is that most PID tuning methods for integrating processes are very aggressive and do not provide a method to adjust the aggressiveness of the closed loop response. If the controller proportional gain (P) is reduced to make the closed loop response less aggressive, the loop is more likely to oscillate! This is quite the opposite result than when this tactic is used on a self-regulating process! Furthermore, there is significant financial benefit by tuning control loops for desired closed loop response to maximize process performance.

Discussion

There are several variations of the integrating response that are commonly found in industrial processes. The 3 most common variations of the integrating response include the following and they will be analyzed in this paper. The step response of each of these types of dynamics is shown in Figure 3.

- Integrating
- Integrating plus dead time
- Integrating plus dead time plus lag

For the common "liquid in, liquid out" level process, the process dynamics typically do not include significant dead time. However, if the level process involves a transport delay of solids material to a bin, the response could include dead time. Boiler drum levels often have an apparent dead time caused by the inverse response of boiler feed water. A dead time could be introduced if the process measure is delayed by the sensor location, for example a temperature measurement in the outflow line of a reactor. If the level controller is cascaded to a flow loop, the closed loop response of the flow loop is included in the level response which would likely result in a lag (small if the flow loop is tuned properly!). The integrating process might include a lag such as that exhibited by a large stirred reactor temperature or a boiler feed water deaerator tank level. These types of complex integrating responses are quite common in industrial processes across all industries.

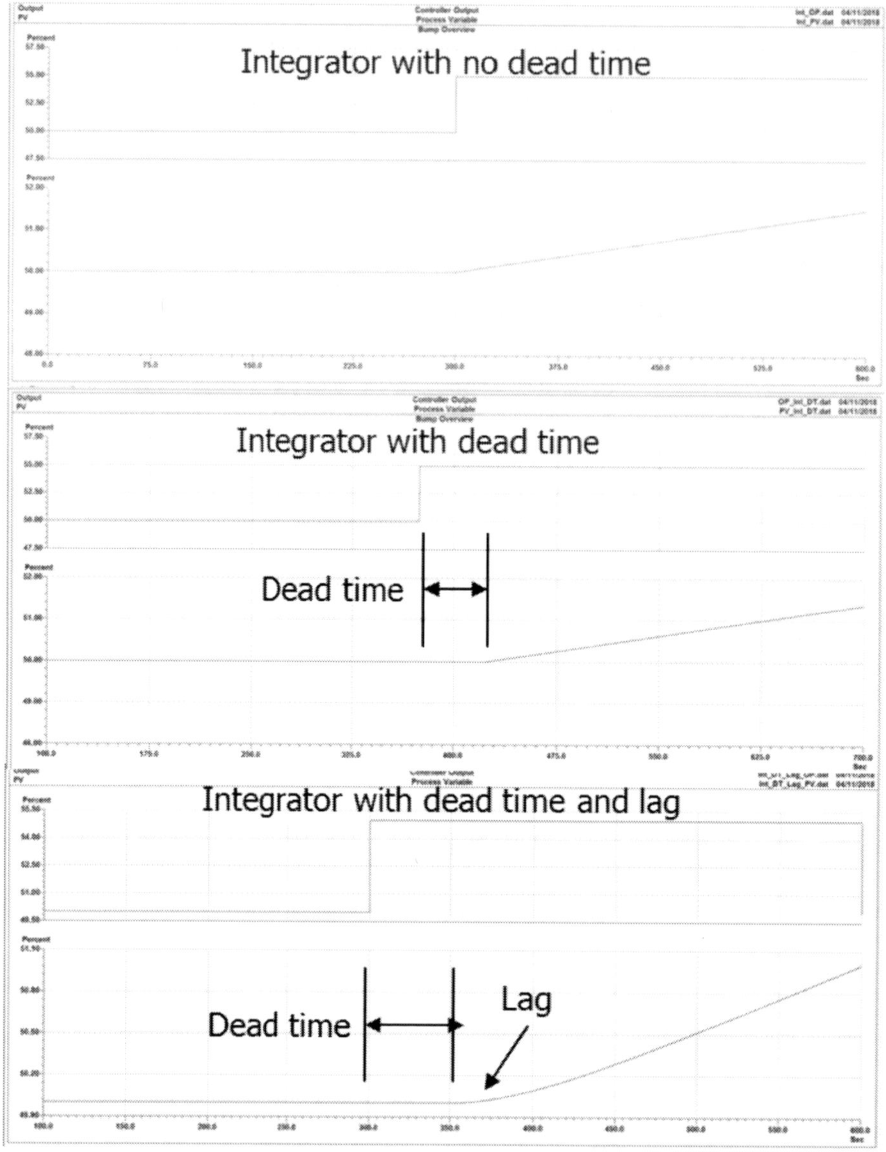

Figure 3 Step Response from Integrating Process with Different Dynamics

The integrating level process with an outlet flow disturbance shown in Figure 4 will be used for the tuning calculations presented in this paper. The 3 most common integrating process responses will be evaluated for this process.

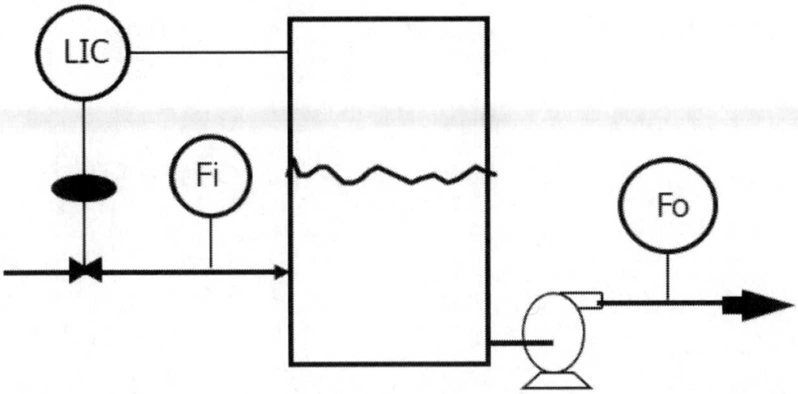

Figure 4 An Integrating Level Process With Load Disturbance Flow, Fo

Figure 5 shows the block diagram of a PID controller applied to an integrating process and a load disturbance, Fd, shown in Figure 4

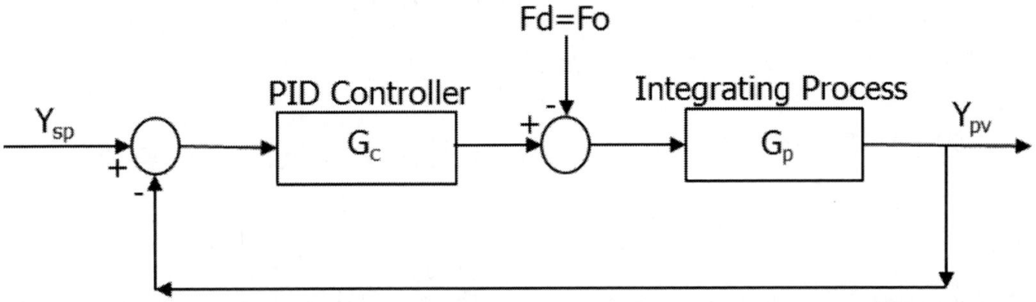

Figure 5 Block Diagram of Integrating Level Process With Load Disturbance Flow

Case 1: Integrating Process with a Proportional Action Only Controller

In case 1, consider an integrating response, without deadtime, and a PID controller with no integral or derivative action. The Laplace transform of the process is shown in equation 1. The Laplace Transform of the "proportional only" PID controller is shown in equation 2. The open loop transfer function, G_{lp}, for this case, is shown in equation 3.

$Gp(s) = \frac{Kp}{s}$ Kp has units of %PV/Second/%OUT Equation (1)

$G_C(s) = Kc$ Kc has units of %OUT/%PV Equation (2)

$Glp(s) = \frac{KcKp}{s}$ Equation (3)

The closed loop transfer function from SP to PV is

$Gsp\ (s) = \frac{\left(\frac{KcKp}{s}\right)}{1+\left(\frac{KcKp}{s}\right)}$ Equation (4)

$Gsp(s) = \frac{KcKp}{s+KcKp}$ Equation (5)

$Gsp(s) = \frac{1}{\left(\frac{1}{KcKp}\right)s+1}$ Equation (6)

$Gsp(s) = \frac{1}{\lambda s+1}$ Equation (7) where $\lambda = \frac{1}{KcKp}$ (units of seconds)

Thus, the closed loop setpoint response is a first order with closed loop time constant λ, lambda. Solving for the proportional only tuning constant, Gain (Kc) results in the following tuning rule.

$Kc = \frac{1}{\lambda Kp}$	Equation (8)

Using an example integrating process with a process gain of 0.001 %PV/Second/%OUT and with 0 dead time, the controller gain Kc for a chosen λ of 500 seconds would be as follows.

$$Kc = \frac{1}{500seconds * \left(0.001\ \frac{\frac{\%PV}{Second}}{\%OUT}\right)} = 2.0\ \%OUT/\%PV$$

Figure 6 shows a simulation of the set point (SP) response of this process and a PID controller with a gain of 2.0 %OUT/%PV (no integral or derivative). Note the closed loop time constant of 500 seconds as chosen and 0% offset between the SP and PV at steady state. Figure 7 shows the simulated response to a step load disturbance of the outflow, Fo.

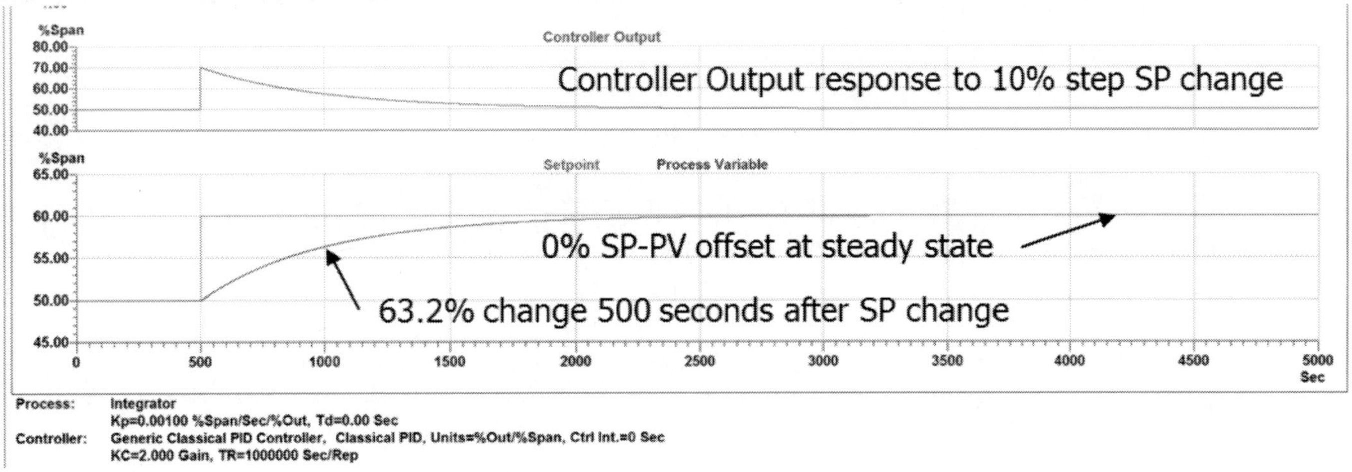

Figure 6 Simulation of Step SP Change for Integrating Process and Proportional-Only PID Controller Tuned for Closed Loop Time Constant, λ, of 500 Seconds

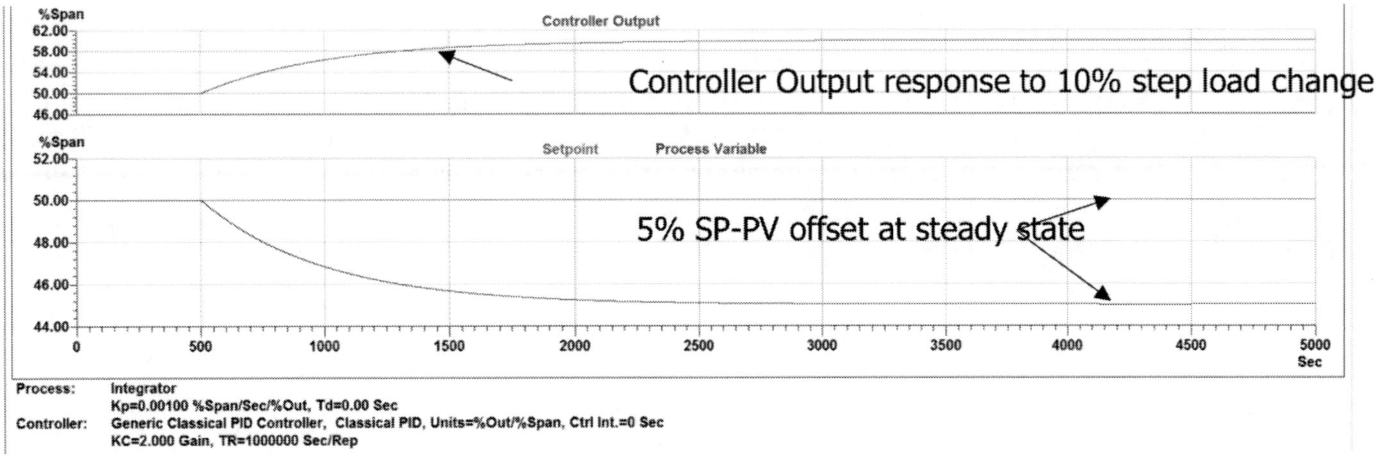

Figure 7 Simulation of Step Load Change for Integrating Process and Proportional-Only PID Controller Tuned for Closed Loop Time Constant, λ, of 500 Seconds

Case 2: Integrating Process with a PI Controller

Note that proportional-only control of a step load disturbance for an integrating process results in an offset between the SP and PV at steady state. This suggests the need for integral action in the controller to eliminate the SP-PV offset. However, if the integral action is too fast, the closed loop response will be oscillatory. The goal is to develop and equation for both the proportional gain and integral action that will provide a non-oscillatory response.

The transfer function, Gc, of a series PID controller with proportional and integral action is:

$$Gc(s) = \left[Kc * \left[1 + \frac{1}{TrS}\right]\right] \qquad \text{Equation (9)}$$

Where Kc has units of %OUT/%PV and Tr has units of Seconds.

The open loop transfer function for the integrating process and series PID controller with proportional and integral action as shown in Figure 5 is:

$$Glp(s) = Gc(s) * Gp(s0 = \left[Kc * \left[1 + \frac{1}{TrS}\right]\right] * \left[\frac{Kp}{S}\right] = \frac{Kc(TrS+1)}{TrS} * \frac{Kp}{S} \qquad \text{Equation (10)}$$

$$Glp(s) = \frac{KcKp(TrS+1)}{TrS^2} \qquad \text{Equation (11)}$$

The root locus for equation (10) and (11) is shown in figure 8. The root locus reveals that with a small reset time Tr (fast integral action) and a low controller gain Kc, the closed loop poles will be very lightly damped and the loop will be likely to oscillate. This is often the case in practice and the typical response to reduce the controller gain for an oscillating loop will make it oscillate even more! To avoid the oscillation due to controller tuning, the tuning should be chosen to locate the closed loop poles all the way around the circle back to the negative real axis where the damping coefficient ζ=1.

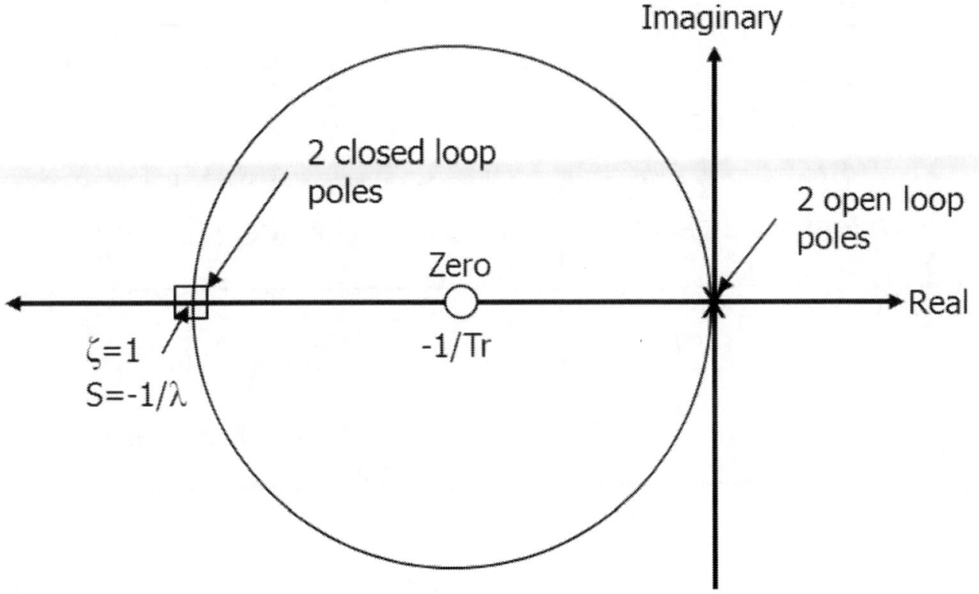

Figure 8 Root Locus diagram for the Open Loop Transfer Function of an Integrating Process and PID Controller with no Derivative Action

Referring to figure 5, the closed loop transfer function G_{ld} from the disturbance Fd=Fo to the process variable Y_{pv} for a series PID controller with proportional an integral action for the integrating process is:

$$Gld(s) = \frac{-Gp(s)}{1+Gc(s)*Gp(s)} = \frac{-\frac{Kp}{s}}{1+\frac{KcKp(TrS+1)}{TrS^2}} = -\frac{\frac{Tr}{KcS}}{\frac{Tr}{KcKp}S^2+TrS+1}$$ Equation (12)

Thus, the characteristic equation for the closed loop control system is:

$$0 = \left(\frac{Tr}{KcKp}S^2 + TrS + 1\right)$$ Equation (13)

The solution to the quadratic equation (13) are the closed loop poles. To have a critically damped closed loop response, the controller tuning parameters Kc and Tr are chosen to have two closed loop poles at -1/λ. When this occurs, the resulting characteristic equation must be as follows:

$$0 = (\lambda S + 1)^2 = \lambda^2 S^2 + 2\lambda S + 1$$ Equation (14)

Comparing this to closed loop characteristic equation (13) results in the following equalities:

$$\frac{Tr}{KcKp} = \lambda^2 \qquad \text{Equation (15)}$$

$$Tr = 2\lambda \qquad \text{Equation (16)}$$

Solving equation 15 for Kc results in the following equations for critically damped tuning of a PID controller with Proportional and Integral action on an integrating process.

$$Tr = 2\lambda \qquad \text{Equation (17)}$$

$$Kc = \frac{4}{Kp*Tr} \qquad \text{Equation (18)}$$

Once again, using an example integrating process with a process gain of 0.001 %PV/Second/%OUT and with 0 dead time, the controller tuning for a chosen λ of 500 seconds would be as follows.

$$Tr = 2\lambda = 2 * 500\, Seconds = 1000\, seconds$$

$$Kc = \frac{4}{Kp * Tr} = \frac{4}{0.001\left(\frac{\frac{\%\text{PV}}{\text{Second}}}{\%\text{OUT}} * (1000\, second)\right)} = 4.0\, \frac{\%OUT}{\%PV}$$

Figure 9 shows a simulation of the closed loop response to a step set point change for an integrating process with a process gain of 0.001%PV/Second/%OUT and 0 second deadtime combined with a Series or Standard PID controller that has a gain of 4.0 %OUT/%PV, 1000 seconds per repeat integral time and 0 derivative. Note that the PV initially reaches the new set point in the time λ after the step set point change. The PV then overshoots the SP by 13.5% and comes back to the SP in time 6* λ after the step SP change. This response is consistent regardless of the size of the step set point change if the output of the PID controller does not hit an upper or lower limit.

Figure 10 shows the simulated response to a step load disturbance of the outflow, Fo. As desired, the PV response is critically damped (no oscillations) and the offset between SP and

PV is reduced to 0 at steady state. Note that the calculated tuning parameters stop, or "arrest" the deviation of the PV from set point in the time λ after the step load disturbance. And, the PV is returned to the setpoint in time 6* λ after the step load disturbance without oscillations. This response is consistent regardless of the size of the step load disturbance change if the output of the PID controller does not hit an upper or lower limit.

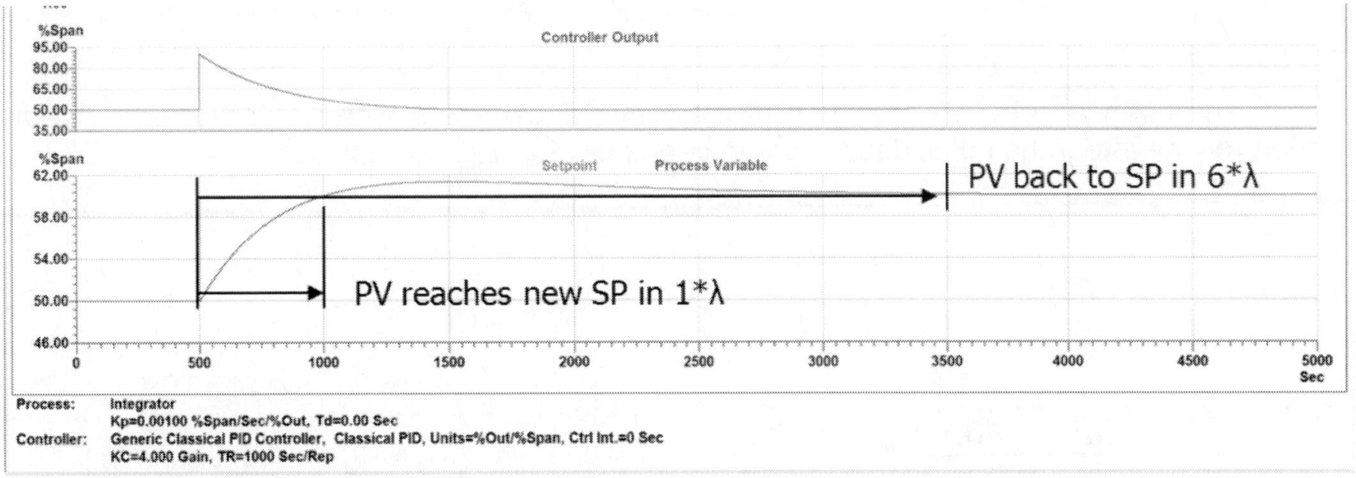

Figure 9 Simulation of Step Set Point Change for Integrating Process and PID Controller With Proportional and Integral Action, Tuned With a λ of 500 Seconds

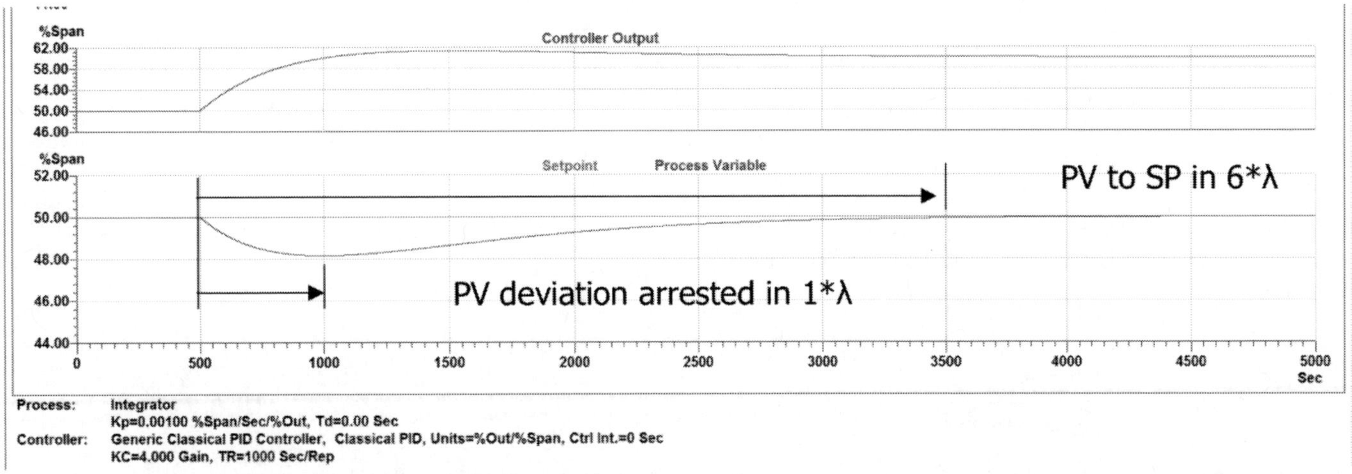

Figure 10 Simulation of Step Load Change for Integrating Process and PID Controller with Proportional and Integral Action Tuned With a λ of 500 Seconds

Case 3: Integrating Plus Deadtime Process with PI Controller

Now, consider the case of an integrating response with dead time, θ, as shown in Figure 3. Referring to Figure 5, the transfer function, Gp, of an integrating response with dead time, θ, is:

$$Gp(s) = \frac{Kp}{S} * e^{-s\theta} \qquad \text{Equation (19)}$$

The open loop transfer function of a series PID controller with proportional an integral action for the integrating plus dead time process shown in Figure 5 is:

$$Glp(s) = Gc(s) * Gp(s) = -\left[Kc * \left[1 + \frac{1}{TrS}\right]\right] * \left[\frac{Kp}{S} * e^{-S\theta}\right] = \frac{Kc(TrS+1)}{TrS} * Kp * \frac{e^{-S\theta}}{S} \qquad \text{Equation (20)}$$

$$Glp(s) = \frac{KcKp(TrS+1)e^{-S\theta}}{TrS^2} \qquad \text{Equation (21)}$$

The set point to PV transfer function is thus:

$$Gsp(s) = \frac{\frac{KcKp(TrS+1)e^{-S\theta}}{TrS^2}}{1+\frac{KcKp(TrS+1)e^{-S\theta}}{TrS^2}} \qquad \text{Equation (22)}$$

This can be simplified to the following:

$$Gsp(s) = \frac{(TrS+1)e^{-S\theta}}{\frac{Tr}{KcKP}*S^2+(TrS+1)*e^{-S\theta}} \qquad \text{Equation (23)}$$

Using the following first order Taylor expansion for dead time:

$$e^{-S\theta} \approx 1 - \theta S = -\theta S + 1 \qquad \text{Equation (24)}$$

Substituting this expansion, the set point transfer function equation 22 is simplified to:

$$Gsp(s) = \frac{\frac{KcKp(TrS+1)(-\theta S+1)}{TrS^2}}{1+\frac{KcKp(TrS+1)(-\theta S+1)}{TrS^2}}$$ Equation (25)

And further simplified:

$$Gsp(s) = \frac{(TrS+1)(-\theta S+1)}{\left(\frac{Tr}{KcKp}-Tr\theta\right)S^2+(Tr-\theta)S+1}$$ Equation (26)

Thus, the characteristic equation for the closed loop control system is:

$$Gsp(s) = \left(\frac{Tr}{KcKp} - Tr\theta\right)S^2 + (Tr - \theta)S + 1$$ Equation (27)

The solution to the quadratic equation (27) are the closed loop poles. To have a critically damped closed loop response, the controller tuning parameters Kc and Tr are chosen to have two closed loop poles at -1/λ. When this occurs, the resulting characteristic equation must be as shown in equation 14, repeated here:

$$0 = (\lambda S + 1)^2 = \lambda^2 S^2 + 2\lambda S + 1$$ Equation (14)

Comparing this to closed loop characteristic equation (27) results in the following equalities:

$$\frac{Tr}{KcKp} - Tr\theta = \lambda^2$$ Equation (28)

$$Tr - \theta = 2\lambda$$ Equation (29)

Solving equation 28 and 29 for Kc and Tr, **results in the following equations for critically damped tuning of a Series or Standard PID controller with Proportional and Integral action on an integrating process with dead time.** Due to the approximation of dead time in the derivation of the tuning equations, lambda, λ, must be greater than the dead time.

Note that one cannot simply choose a small lambda, λ, for an integrating plus dead time and achieve the same closed loop performance as with an integrating process with 0 deadtime. The equations are valid for $\lambda > \theta$, and preferably $\lambda > 3*\theta$ to provide reasonable margin of stability. In many cases, the lambda chosen for best process performance could be very large,

>>θ. An example of this is a level controller, on a tank whose capacity is used to absorb process variability, to reduce variability in the manipulated flow and downstream process.

$Tr = 2\lambda + \theta$	where λ>θ	Equation (30)
$Kc = \frac{Tr}{Kp*(\lambda+\theta)^2}$	where λ>θ	Equation (31)

Using an example integrating process with a process gain of 0.001 %PV/Second/%OUT and with 50 seconds dead time, the controller tuning for a chosen λ of 500 seconds would be as follows.

$$Tr = 2\lambda + \theta = (2 * 500 + 50)\,Seconds = 1050\,seconds$$

$$Kc = \frac{Tr}{Kp * (\lambda + \theta)^2} = \frac{1050\,seconds}{0.001\,\frac{\frac{\%\text{PV}}{\text{Second}}}{\%\text{OUT}} * \left((500 + 50)seconds\right)^2} = 3.47\,\frac{\%OUT}{\%PV}$$

Figure 11 shows a simulation of the closed loop response to a step set point change for an integrating process with a process gain of 0.001%PV/Second/%OUT and 50 second deadtime combined with a Series PID controller that has a gain of 3.47 %OUT/%PV, 1050 seconds per repeat integral time and 0 derivative. Note that the PV initially reaches the new set point in the time λ after the step set point change. The PV then overshoots the SP by 13.5% and comes back to the SP in time 6* λ after the step SP change. This response is consistent regardless of the size of the step set point change if the output of the PID controller does not hit an upper or lower limit.

Figure 12 shows the simulated response to a step load disturbance of the outflow, Fo. As desired, the PV response is critically damped (no oscillations) and the offset between SP and PV is reduced to 0 at steady state. Note that these tuning values stops, or "arrests" the deviation of the PV from set point in the time λ after the step load disturbance. And, the PV is returned to the setpoint in time 6* λ after the step load disturbance without oscillations. This response is consistent regardless of the size of the step load disturbance change if the output of the PID controller does not hit an upper or lower limit.

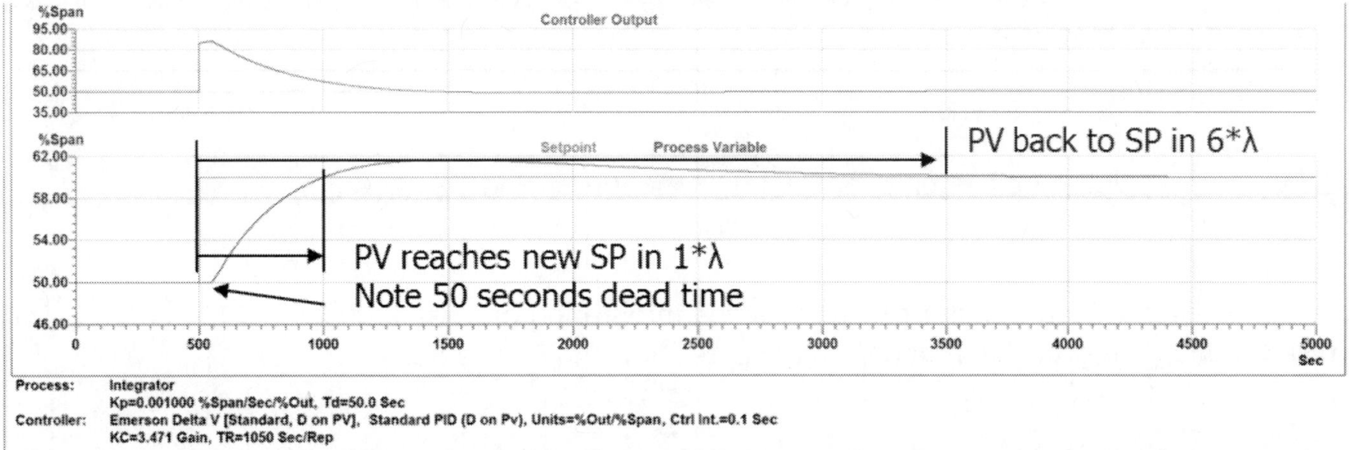

Figure 11 Simulation of Step Set Point Change for Integrating Process with Dead Time and PID Controller With Proportional and Integral Action, Tuned With a λ of 500 Seconds

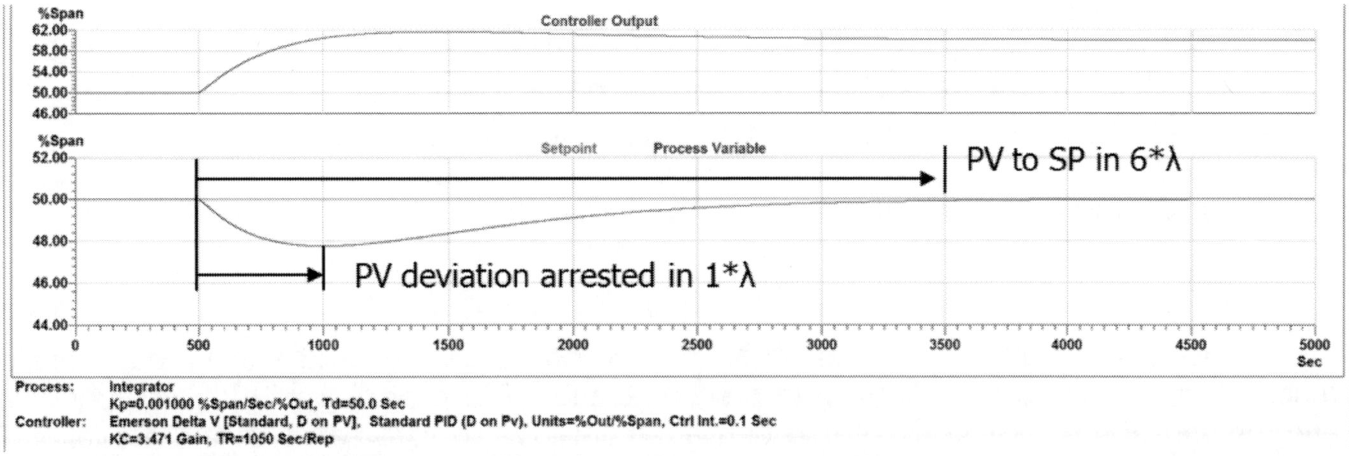

Figure 12 Simulation of Step Load Change for Integrating Process with Dead Time and PID Controller With Proportional and Integral Action Tuned With a λ of 500 Seconds

Case 4: Integrating Plus Deadtime and Lag Process with PID Controller

Now, consider the case of an integrating response with dead time, θ, plus lag, τ, as shown in Figure 3. Referring to Figure 5, the transfer function, Gp, of an integrating response with dead time, θ, plus lag, τ, is:

$$Gp(s) = \frac{Kp}{S} * \frac{1}{\tau S+1} * e^{-s\theta}$$ Equation (32)

The transfer function, Gc, of a Series (aka Classical) PID controller with proportional, integral and derivative action is:

$$Gc(S) = Kc[TdS + 1] * \left[1 + \frac{1}{TrS}\right]$$ Equation (33)

Where Kc has units of %OUT/%PV and Tr and Td have units of Seconds.

The open loop transfer function for the integrating process and series PID controller with proportional, integral action and derivative as shown in Figure 5 is:

$$Glp(s) = Gc(s) * Gp(s0 = [Kc[TdS + 1] * \left[1 + \frac{1}{TrS}\right]] * [\frac{Kp}{S} * \frac{1}{\tau S+1} * e^{-s\theta}]$$ Equation (34)

Note that if the derivative time, Td, of the series PID controller is set equal to the process lag, τ, the two terms cancel each other and the open loop transfer function is:

$$Glp(s) = Kc * \left[1 + \frac{1}{TrS}\right] * \left[\frac{Kp}{S} * e^{-s\theta}\right] = \frac{KcKp*(TrS+1)*e^{-s\theta}}{TrS^2}$$ Equation (35)

Note that equation (35) is identical to equation (21) which is the open loop transfer function for an integrating process with dead time and a PID controller with only proportional and integral action. Thus, by setting derivative time, Td, of the Series (aka Classical) PID controller equal to the process lag, τ, the controller gain, KC, and integral time, Tr, can be calculated in the same way as was done for an integrating process with dead time. Due to the approximation of dead time in the derivation of the tuning equations, lambda, λ, must be greater than the dead time.

Again, note that one cannot simply choose a small lambda, λ, and achieve the same closed loop performance as with a integrating process with 0 deadtime. The equations are valid for $\lambda > \theta$, and preferably $\lambda > 3*\theta$ to provide reasonable margin of stability. In many cases, the lambda chosen for best process performance could be very large, $>>\theta$.

Equations (36), (37) and (38) provide the tuning equations for a Series PID controller for an integrating process with dead time.

$Td = \tau$		Equation (36)
$Tr = 2\lambda + \theta$	where $\lambda>\theta$	Equation (37)
$Kc = \frac{Tr}{Kp*(\lambda+\theta)^2}$	where $\lambda>\theta$	Equation (38)

Using an example integrating process with a process gain of 0.001 %PV/Second/%OUT and with 50 seconds dead time, the controller tuning for a chosen λ of 500 seconds would be as follows.

$$Td = \tau = 50\ seconds$$

$$Tr = 2\lambda + \theta = (2 * 500 + 50)\ Seconds = 1050\ seconds$$

$$Kc = \frac{Tr}{Kp * (\lambda + \theta)^2} = \frac{1050\ seconds}{0.001\ \frac{\frac{\%\mathrm{PV}}{\mathrm{Second}}}{\%\mathrm{OUT}} * \left((500 + 50)seconds\right)^2} = 3.47\ \frac{\%OUT}{\%PV}$$

Figure 13 shows a simulation of the closed loop response to a step set point change for an integrating process with a process gain of 0.001%PV/Second/%OUT and 50 second deadtime combined with a Series PID controller that has a gain of 3.47 %OUT/%PV, 1050 seconds per repeat integral time and 50 seconds derivative time. Note that the PV initially reaches the new set point in the time λ after the step set point change. The PV then overshoots the SP by about 20% and comes back to the SP in time 6* λ after the step SP change. This response is consistent regardless of the size of the step set point change if the output of the PID controller does not hit an upper or lower limit.

Figure 14 shows the simulated response to a step load disturbance of the outflow, Fo. As desired, the PV response is critically damped (no oscillations) and the offset between SP and PV is reduced to 0 at steady state. Note that these tuning values stops, or "arrests" the deviation of the PV from set point in the time λ after the step load disturbance. And, the PV is returned to the setpoint in time 6* λ after the step load disturbance without oscillations. This response is consistent regardless of the size of the step load disturbance change if the output of the PID controller does not hit an upper or lower limit.

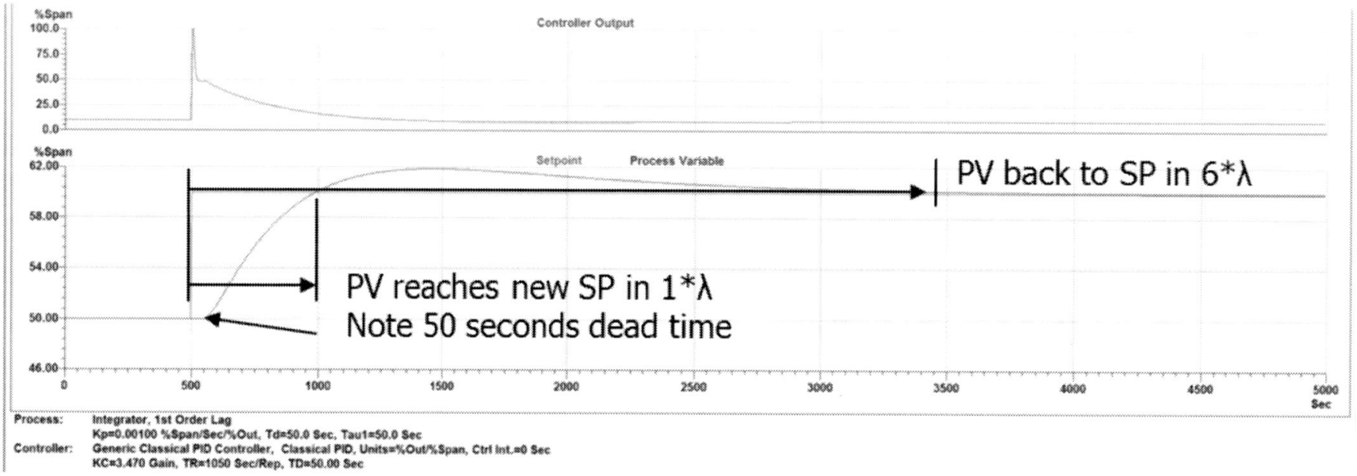

Figure 13 Simulation of Step Set Point Change for Integrating Process with Dead Time and Lag and a PID Controller With Proportional, Integral and Derivative Action, Tuned With a λ of 500 Seconds

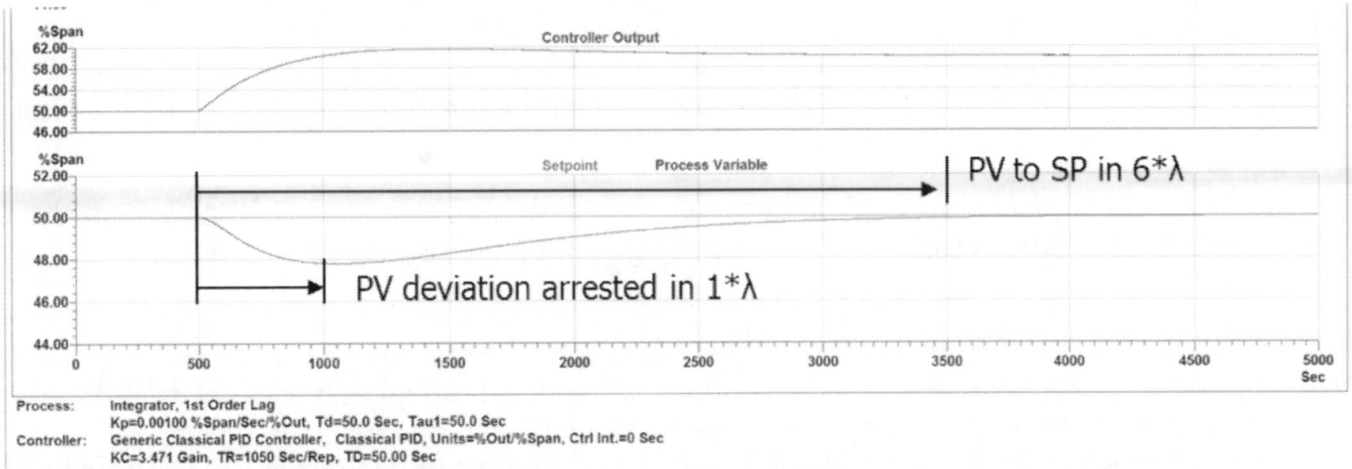

Figure 14 Simulation of Step Load Change for Integrating Process with Dead Time and Lag and a PID Controller With Proportional, Integral and Derivative Action, Tuned With a λ of 500 Seconds

Conversion of Tuning from Series to Standard PID

The tuning equations presented in this paper are based on a Series Implementation of the PID control algorithm. If the derivative time is set to 0 (no derivative action), the Series and Standard PID algorithms are identical. If the derivative time is not 0, the Series tuning parameters can be converted for the Standard PID algorithm.

$$Kc_{std} = Kc * \frac{Tr+Td}{Tr}$$ Equation (39)

$$Tr_{std} = Tr + Td$$ Equation (40)

$$Td_{std} = \frac{Tr*Td}{Tr+Td}$$ Equation (41)

Summary

Tuning PID controllers for integrating process is often counter-intuitive. There are few tuning methods devoted to integrating process and most methods do not allow the choice of a closed loop speed of response. This feature is very useful when tuning the loop to maximize process performance vs. loop performance. This task is further complicated when the integrating response includes additional dynamics such dead time and lags. The tuning calculations presented in this paper are based in Internal Model Control theory where the tuning parameters are calculated to achieve the desired closed loop response. Note that tuning equations are based on a chosen transform for dead time which can impact the results but well within the range of variation of process models in the industrial process industry. The tuning equations have been used on thousands of control loops providing significant financial benefits.

There are additional cases of integrating process dynamics and tuning options which have not been explored in this paper. These cases can be discussed with the author if desired.

The tuning equations developed in this paper for a Series (aka Classical) PID Form are repeated here for convenience. There are many variations of implementation of the PID control function as well as many options. These variations and options may impact the suitability of these equations for a specific application. Good engineering practices should be used to verify the applicability of these equations for the application.

Process: Integrator

P only Controller: $Kc = \frac{1}{\lambda Kp}$ Equation (8)

PI Controller: $Tr = 2\lambda$ Equation (17)

(Series or Standard PID) $Kc = \frac{4}{Kp * Tr}$ Equation (18)

Process: Integrator plus dead time

PI Controller: $Tr = 2\lambda + \theta$ where λ>θ Equation (30)

(Series or Standard PID) $Kc = \frac{Tr}{Kp * (\lambda + \theta)^2}$ where λ>θ Equation (31)

Process: Integrator plus dead time and lag

Series PID Controller: $Td = \tau$ Equation (36)

$Tr = 2\lambda + \theta$ where λ>θ Equation (37)

$Kc = \frac{Tr}{Kp * (\lambda + \theta)^2}$ where λ>θ Equation (38)

512681 Tune Your PID Loops Please!

Abstract:

"Who looks after PID tuning at your site?" This is a question I often ask many professionals. The answers vary from a definitive deparment to "don't know" or "maintenance department", "instrumentation department" etc. The replies almost always suggest to me that there is not enough focus or attention paid to this important but generally forgotton layer of automation.

PID (Proportional-Integral-Derivative) controllers still form the backbone of automatic control systems in refineries, (petro) chemical installations, power plants and numerous other processes. The success of PID schemes lies in the simplicity of the implementation, reliability of the hardware and the stabilising effect on processes. It is however necessary to enter values for the "P", "I" and "D", referred to as tuning parameters, to make this PID controller function in a stable and efficient manner.

Significant benefits can be achieved by optimising the performance of PID loops, especially interacting loops. This presentation explores optimisation possibilities for process industries with some case studies on PID tuning and introduces a new paradign on how to undertake a PID tuning project at your facility.

511872 pH and Other Non-Linear Control Applications

Abstract:

The control of pH has many very important applications in industry. The non-linearity of pH control can make it extremely challenging using conventional PID control. This presentation examines non-linear control techniques to improve feedback pH control with applications to other non-linear control challenges. We will discuss techniques including several types of variable gains, linearization with ion concentration and titration curves. Change of variable will also be discussed. The advantages and disadvantages of each technique will be presented so that you can determine how to choose between the solutions to optimize performance in the intended application.

Cooperative Distributed Control: Model Predictive Control for the Industrial Internet of Things

AIChE 2018 Spring Meeting and 14th Global Congress on Process Safety
Session: Process Control and Optimization Developments - Wednesday, April 25, 2018

Dr Patrick Thorpe CEng FIChemE, patrick.thorpe@spirocontrol.com

Summary

Cooperative distributed model predictive control is receiving significant attention as a major next generation Model Predictive Control (MPC) technology and key component in unlocking the potential of the Industrial Internet of Things (IIoT). Cooperative distributed control enables a distributed network of cooperating controllers to collaborate towards a common objective without use of a centralized coordinator or optimizer. In addition to presenting a simple, modular route to plant-wide control and optimization, the technology offers a general approach to distributed MPC application within an IIoT framework.

Introduction

Model Predictive Control has been the de-facto advanced control technology used in the oil and gas sector to optimise refineries and petrochemical plants for several decades. Saving many millions of dollars per year in energy and increasing profits through improved plant throughput and product yields. However, due to its perceived high cost, complexity and restricted flexibility, MPC has seen limited adoption outside specific application areas. The inflexibility of existing MPC technology is particularly apparent when considering its application to large distributed but interdependent facilities. A further limiting factor has been the inability to incorporate future forecast of a changing environment into the control strategy. (Rawlings and Mayne 2013 provide a comprehensive overview of MPC theory).

Cooperative distributed control is an MPC based technology that overcomes some of these limitations. The solution enables a distributed network of controllers to collaborate towards a common objective without use of a centralized coordinator or optimizer, obeying all constraints of the network and utilizing available future planning forecasts such as production changes or environmental conditions. By eliminating the centralized component, the technology can work within complex and geographically distributed control structures. Crucially the technology lends itself to a plug and play architecture allowing a new sub-system model, variable, or hardware device to be added to and automatically assimilated into the cooperative network with relative ease. The embedded capability of cooperative control enables it to be used for applications that are not readily addressed by existing solutions including compressors, fired heaters, steam networks etc. In addition, the technology is designed to address a new range of control problems including gas distribution networks, building control, power grid load management etc. (See, for example, Stewart 2010, Razzanelli 2016, Shaoyuan 2015, Christofides 2013 for discussions of Cooperative Distributed MPC).

Outline of method

The proposed technology is designed to be integrated as a function block within an existing DCS controller, or on an edge device such as a Programmable Automation Controller (PAC) or Programmable Logic Controller (PLC). The code has been optimized for efficient execution and has a low memory footprint and has been developed in industry standard IEC 61131-3 allowing porting to many industrial controllers with minimal customization.

The unique architecture allows a multivariable control function to be distributed across many cooperating function blocks. This facilitates load distribution and ensures that peak CPU load is maintained within required limits. A large multivariable control solution can be deployed by interconnecting these simple function blocks. Cooperative controllers can also be distributed across multiple control processors.

What follows is a basic outline of the theory that facilitates cooperation between controllers. Consider a typical dynamic optimization function for a simple plant with two subsystems and two manipulated variables (u1 and u2):

$$V(x_1(0), x_2(0), u_1, u_2) = \sum_{i=1}^{N-1} \left[\frac{1}{2}[x_i, u_i]_i^T H_i [x_i, u_i]_i + f_i^T [x_i, u_i]_i\right] + [x_i, u_i]_N^T H_N [x_i, u_i]_N$$

$$x_{i+1} = Ax_i + B_1 u_i + B_2 u_2$$

$$P_i[x_i, u_i] \leq b_i$$

Each subsystem solves the above objective function for a known and fixed value of other subsystem input sequences. The process is repeated over a series of iterative steps.

1. (System –1) Fix u2 and solve with respect to u1

$$Set\ u_2 = u_2^p: \ \min_{u1} V(x_1(0), x_2(0), u_1, u_2) \rightarrow (u_1^*, u_2^p)$$

2. (System –2) Fix u1 and solve with respect to u2

$$Set\ u_1 = u_1^p: \ \min_{u2} V(x_1(0), x_2(0), u_1, u_2) \rightarrow (u_2^*, u_1^p)$$

3. Combine the results and make a step

$$u_i^{p+1} = \alpha(u_i^p) + (1-\alpha)(u_i^*)$$

The key design characteristic of cooperative control is that it solves the multivariable optimization without centralized coordination – each controller can operate autonomously and stably in the case of communication loss. Furthermore, cooperative iterations can terminate early (before full convergence) without compromising closed loop stability. With sufficient cooperative iterations the solution will converge to the global optimum and is equivalent to a single system wide multivariable controller.

There is significant flexibility in how the solution is deployed. Cooperative control *agents* can be separate multi-variable control applications configured on the same device or distributed over many *edge* devices. The edge devices may be geographically dispersed across a network.

Example 1: Plant-wide control and optimisation of an ethylene process

One of the use cases for the technology is to improve on current methods for controlling a large integrated process such as an ethylene plant or gas plant. Ethylene is commonly produced by steam cracking a petrochemical feedstock such as ethane. Figure-1 shows a simplified ethylene process schematic comprising a single steam cracking furnace, compressor, demethaniser column, expander, deethaniser column and C2-splitter column.

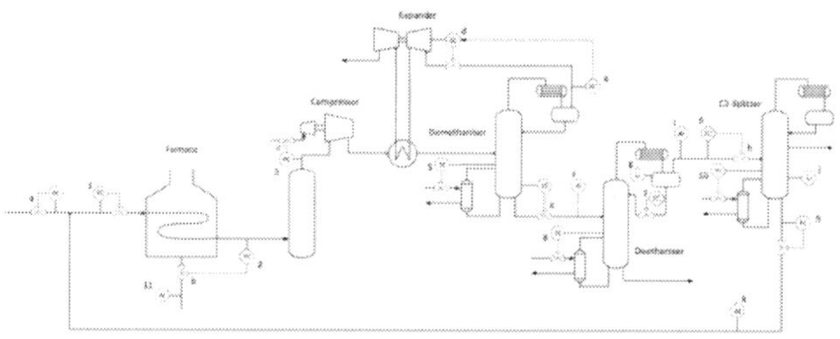

Fig 1. Simplified Ethylene Process Schematic

Each section of the ethylene process is considered a separate unit operation. It is commonplace for each unit operation to be controlled using a series of independent Proportional Integral Derivative (PID) control loops or model predictive controllers. However, due to the existence of strong interactions between unit operations in the steam cracking process, individual controllers are unable to achieve stable or optimal control. To overcome these challenges, the process may be controlled by a single monolithic model predictive controller or optimizer encompassing the complete process. Whilst this strategy can deliver significant benefits, there are substantial limitations relating to the complexity and engineering effort required to develop and maintain such an application.

In the proposed approach, separate control agents can be configured for each process section (steam cracking furnace, quench column, distillation columns etc.), thereby maintaining the simplicity of distributed control. However, the control agents are designed to operate cooperatively such that in addition to control of the local process section, each agent is also aware of global constraints and interactions between process sections. By operating cooperatively, control agents can approach a global optimum and ensure stability of each individual process section.

Example 2: Gas distribution network

Figure 2 shows a typical supply chain for the provision of ethane feed stock to a gas fed ethylene plant. Natural gas, which may be associated gas extracted with crude oil or non-associated gas, is fed to the gas plant for extraction of valuable products including ethane and higher hydrocarbons from the less valuable sales gas. NGL product is sent as a mixture of ethane, propane and butane by pipeline to the NGL plant where it is separated into its constituent components. The lightest component, ethane, is extracted in the first NGL column, which is the deethaniser, and sent as feed to the ethylene plant. It should be noted that each of the process plants (gas plant, NGL plant and ethylene plant) is a separate entity, with separate control facilities and process operators. Often the gas plant is geographically separated from the NGL plant with a pipeline supplying the NGL feed over large distances.

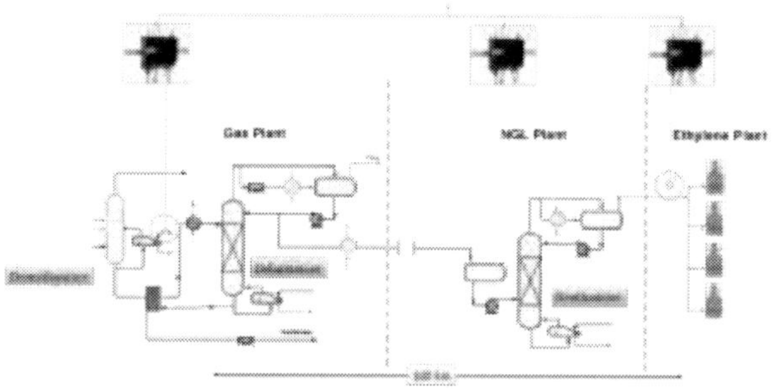

Fig 2. Ethylene Plant Feed Supply

Due to the geographic and operational separation, each plant is controlled independently with information shared only at the planning level. This approach ignores the dynamic interactions between the separate processes. For example, the control temperature of the gas plant demethaniser has a strong influence on the total ethane supply that is available to the ethylene plant. In the case of an operating disruption or feed reduction on the ethylene plant, the surplus ethane must be flared. In this situation it is far more economic to raise the gas plant demethaniser temperature, allowing the surplus ethane to be extracted with the local sales gas.

A cooperative distributed control solution covering the gas plant, NGL and ethylene plant with a shared objective would allow controllers on each plant to anticipate feed disruptions and minimize flaring while also maximizing the available ethane feed to the ethylene plant. This has the potential for significant financial benefits and a reduction in environmental emissions. The benefits in this case would be much higher than those derived from a steady state optimizer or planning model because of the need to respond quickly, in real-time, to shifting operating conditions.

Example 3: Utilities plant steam network

A typical process plant steam system may include a range of different steam generators connected to the network, all capable of producing steam at the required grade. However, different generators operate at different efficiencies. For example, there may be highly efficient Combined Heat and Power (CHP) units, less efficient gas fired boilers and finally oil-fired boilers with lower efficiency and higher emissions. The objective is to generate the required quantity of steam at the lowest total energy and within emissions limits.

It would be possible to optimize the overall steam network with a centralized optimizer that looked at the efficiency and generating capacity of each unit and coordinated the overall operation to maintain the required steam header pressure. But this centralized optimizer could be complex to develop and maintain and is rendered inoperable if new generators are added to the network.

With cooperative distributed control, each generator is treated as a separate sub-system with a local control agent. Each local control agent collaborates with others in the network so that collectively they agree on the best way to generate the required amount of steam. If there is a disruption to the steam network or one generator drops out, the remaining agents quickly re-coordinate to maintain the overall system. When a new steam generator is added it can be quickly assimilated into the network.

Example 4: Building control

Buildings consume more than 40% of the total primary energy resources throughout the world with an estimated 82% unrealized energy efficiency potential (IEA 2014). MPC is gaining popularity as an effective approach to building climate control given that it is a typical multi-variable constrained control problem. The main aim of MPC when applied to building management is to reduce the energy consumption, considering occupant comfort expectations.

Almost all buildings are controlled by independent zone controllers in a distributed architecture. However, this approach neglects the thermal coupling that exists between adjacent rooms, through internal walls, floors and door openings. It is for this reason that decentralized control schemes return a sub-optimal solution compared to centralized MPC. Moroşan et al. (2010) compared a conventional baseline controller with a decentralized and centralized MPC scheme, which achieved energy savings of 5.5% and 13.4%, respectively.

The limitation of centralized MPC is that as the system becomes more complex and spatially distributed, a significant amount of computing time and data transfer are required, leading to communication overhead. In case of communication breakdown, the controller is no longer able to compute the control signals. Consequently, for large buildings, the prevailing view is that despite the performance benefits of centralized control, for reliability issues it is still advisable to deploy a distributed architecture.

Cooperative zone controllers represent the best of both worlds accounting for thermal interactions and building wide constraints, enabling improved control with lower total energy. Additionally, cooperative controllers can incorporate common, building wide objectives such as a limit on the total power demand which may be particularly relevant when power is supplied by a renewable resource such as solar. The controllers can also take account of future planning data such as room occupancy scheduling, weather forecast and utility pricing tariffs. A simple, modular construction based on relatively simple, low cost hardware means cooperative distributed control presents significant advantages over a centralized building management system.

Example 5: Microgrids

A microgrid is a grouping of distributed energy sources with control capabilities that can work while connected to the main power network or can operate autonomously as dictated by demand. Microgrids range in size from large industrial or community systems, down to small localized systems.

Cooperative distributed control can be applied to perform optimal load management of a microgrid. Optimal load management can be achieved by implementing a moving time horizon window to plan the optimal grid strategy over a future time-period of up to 24-hours. Input data for calculating the optimal grid strategy can include forecast data on variables such as ambient temperature, solar intensity, wind speeds, expected load profile and electricity supply tariffs. In this way the system ensures the best use of available resources including energy storage devices whilst accounting for constraints such as total grid load and equipment operating limits. As the technology derives from closed-loop model predictive control, it means that the plan can be continuously updated in real-time based on a comparison of actual vs forecast data.

Figure 3 illustrates the power flows in a simple microgrid simulation comprising solar panels, conventional diesel generator, battery storage, load, and utility grid connection. The battery can store power when the grid price is low or when renewable output is high and can be made to discharge when the grid price is high or when the renewable output is low. Buying and selling with the utility grid is available based on the applicable tariff. There is an agreed cap on peak power consumed from the grid (12 kW). A conventional diesel generator is available to manage peak load. Each microgrid component is controlled by a separate cooperative control agent with agents cooperating to optimize the common objectives and constraints.

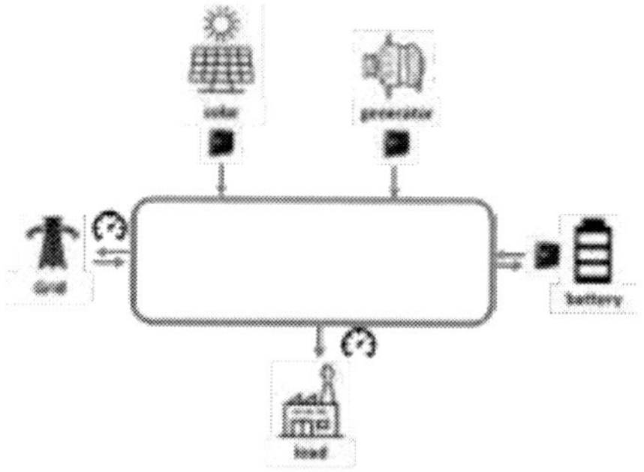

Fig 3. Simple microgrid illustration

Figure 4 shows typical forecast data for solar power generation, load profile and pricing tariff over a 24-hour period.

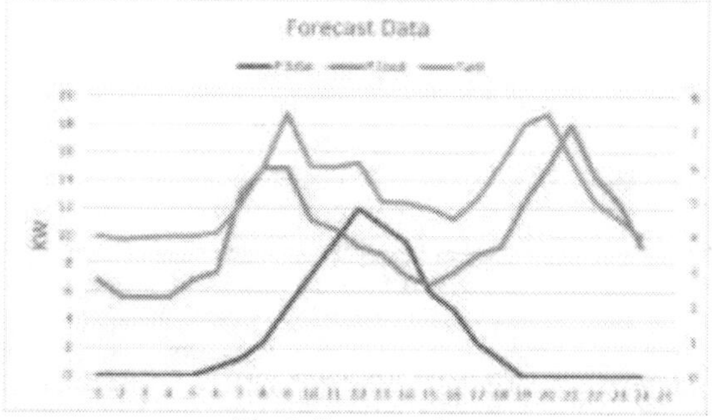

Fig 4. Forecast data

Cooperative controllers on the generator and battery can use the available forecast data, together with metered data throughout the day, to optimize overall power use. The benefit is shown in comparison to a reactive case where battery and generator use are determined by considering current load requirements, generated solar power, and spot tariff data without forecasting.

Figures 5 and 6 show the power drawn from the utility grid (positive values), power sold to the utility grid (negative values), battery state of charge (SOC) and the total normalized cost per operating hour. By forecasting the anticipated daily load and tariffs, the cooperative controllers make better use of the available battery storage capacity and conventional generator to lower the total electricity cost compared to the reactive case.

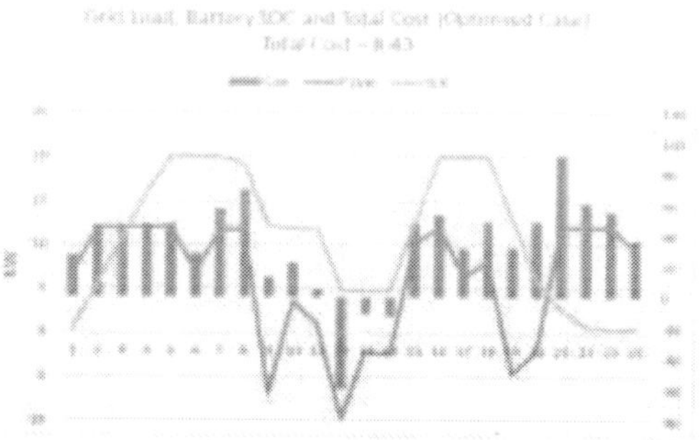

Fig 5. Optimized case

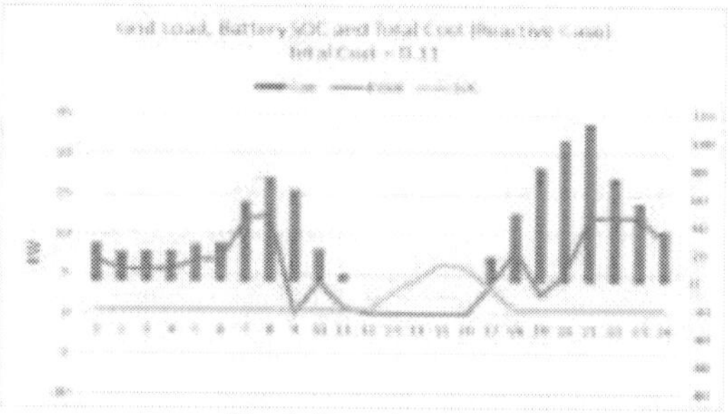

Fig 6. Reactive case

Conclusion

The Industrial Internet of Things (IIoT) offers unprecedented opportunities for industrial organizations to improve their operational efficiency. Hosted by IIoT enabled, open, secure, and interoperable system architectures, cooperative distributed control technology enables multiple interconnected and cooperating edge controllers to optimize an entire building, industrial process or utility network. This technology has the potential to harness the power of IIoT connectivity to transform process control and to deliver significant operational benefits. Additionally, it has the advantage of a simple, modular design and utilizes existing control standards and infrastructure.

References

Christofides, Scattolini, Muñoz de la Peña and Liu, Distributed model predictive control: A tutorial review and future research directions, Computers and Chemical Engineering, Volume 51, April 2013.

International Energy Agency (IEA) Capturing the Multiple Benefits of Energy Efficiency, 2014.

Moroşan, Bourdais, Dumur and Buisson. Building temperature regulation using a distributed model predictive control. Energy Build. 42 (2010) 1445-1452.

Razzanelli and Pannocchia, Parsimonious Cooperative Distributed MPC for Tracking Piece-Wise Constant Setpoints, IFAC Papers On Line, Volume 49, Issue 7, 2016, Pages 520-525

Rawlings and Mayne, Model Predictive Control: Theory and Design, Nob Hill Publishing, 2013.

Shaoyuan and Zheng, Distributed Model Predictive Control for Plant-Wide Systems, Wiley, 2015.

Stewart, Venkat, Rawlings, Wright and Pannocchia, Cooperative distributed model predictive control, Systems and Control Letters 59 (2010) 460-469.

513904 Automation of Knowledge Work in Advanced Process Control

Abstract:

Multivariable Predictive Control (MPC) has been an active area in both academic research and industrial applications for over 25 years. It still creates a tremendous value in the process industries by optimizing operations by increasing throughput, maximizing yields and reducing energy consumption. Although many advanced algorithms have been developed over the years in the area of model identification optimization and control, the methodology for the deployment of the MPC projects has not changed significantly and many of the MPC project steps are still manual and tedious, requiring significant effort and time. As the chemical processes and operation conditions continuously change, the process models need to be frequently re-identified and the controller tuning parameters need to be re-tuned in order to sustain the economic benefits of the MPC application.

Automation of knowledge work has been identified as one of the most impactful and disruptive initiatives with similar economic impact to the Industrial Internet of Things (IIOT). In this presentation we will discuss how automation of knowledge work applies to MPC in order to automate and simplify complex analysis and workflows and thus reduce barriers to deployment. More specifically we will look into automation of knowledge work for online Model Quality Analysis, Seed Model Generation, Automatic Data Slicing, Automatic Constrained Model Identification, Colinearity analysis and repair etc.

One particular area that we will focus on is automating the complex process of tuning the steady-state MPC optimizer via a Sequential Multi-Objective Optimization (SMO) that enables the user to simply specify the operational objectives of the process. An industrial example will be presented to demonstrate that via this approach the LP tuning remains consistent and does not need to be revisited when the controller models are updated.

AUTHOR INDEX

AIChE
120 Wall Street, FL 23
New York, NY 10005-4020

ISBN 978-1-5108-6447-4

CURRAN ASSOCIATES INC.
proceedings.com